Asset Management Insights

Phases, Practices, and Values

Asset Management Insights

Phases, Practices, and Values

Celso de Azevedo, Ph.D.

INDUSTRIAL PRESS, INC.

Industrial Press, Inc.
32 Haviland Street, Suite 3
South Norwalk, Connecticut 06854
Phone: 203-956-5593
Toll-Free in USA: 888-528-7852
Email: info@industrialpress.com

Author: Celso de Azevedo, Ph.D.
Title: Asset Management Insights: Phases, Practices, and Value
Library of Congress Control Number: 2019936357

Printed in the United States of America.

ISBN (print)	978-0-8311-3644-4
ISBN (ePUB)	978-0-8311-9532-8
ISBN (eMOBI)	978-0-8311-9533-5
ISBN (ePDF)	978-0-8311-9530-4

Publisher/Editorial Director: Judy Bass
Copy Editor: Janice Gold
Compositor: Patricia Wallenburg, TypeWriting
Proofreader/Indexer: Claire Splan

books.industrialpress.com
ebooks.industrialpress.com

10 9 8 7 6 5 4 3 2 1

To the memory of my brother, Wilton,

with whom I shared the most beautiful learning:

that of life.

Contents

PART III

The Operational Life

PART IV

The End-of-Life

Foreword

The Emergence of a Necessity

To produce a historical account of our discipline's genesis would be an exercise both tedious and, in the end, only sparsely instructive. What strikes us as more pressing, considering the shortfalls, the approximations, and the genuine mistakes made on a regular basis by some of those who claim to know our subject's intricate workings, is to come back to the deeply rooted causes of its emergence in order to demonstrate that its development was inscribed in a rationale of *necessity*.

In the 1980s, as in other turning moments in history, the industrial and infrastructural worlds experienced a paradigm shift, thus setting the scene for a positive transformation of our relation to the management of assets, namely the means (physical or not) through which an organization seeks to extract value from the assets' operation throughout their life cycles. By illustrating this progression, we hope to shed a light on the foundations and the inherent dynamics and logics of Asset Management, as well as to respond to the questions and criticisms expressed by those who claim, against all reason, to view it as nothing more than a superfluous appendix of industrial maintenance or, for that matter, of any operational discipline that plays a part in the challenges of Asset Management without embodying its totality.

Similarly, we wish to prove wrong those who, failing to understand our discipline's mechanisms, dogmatize it to the point of rigidity; the very self-proclaimed experts who, when they are not busy dogmatizing Asset Management, feel entitled to discuss it even though they can only boast a superficial or partial knowledge of the matter. In other words, we intend to provide a positive vision of the management of assets, founded on a true conceptual rigor but exemplified by experience.

Needless to say, this demonstration cannot claim to offer a somewhat utopian character of *certainty*. I myself have been so lucky as to pursue just about the entirety of my career in this discipline's emergence and momentum; therefore, my perspective is eminently initiatory, and simultaneously covers a practice spread over several decades, a conceptual stack validated by empiricism, and the occasional trials and errors of the scientific thought.

Furthermore, it goes without saying that the history of scientific thought should not be devised as a straight axis, whizzing through the ages without twisting and turning. On the contrary, we know that science advances through experience, and therefore through trials and failures—progress deriving, *in fine*, from our right to make mistakes. This road towards progress is therefore characterized by the systematic confrontation of "old" questions to innovative and crystallizing solutions, themselves made available by the frequent theoretical achievements that accompany the process. It is therefore obvious that a more performant system can only be designed by learning from the dysfunctions of one previously experienced—which isn't to say that we should reject as a whole every particular element of an anterior model, but rather that we should learn to draw the consequences from past failures, as soon as novel and superior ideas have been demonstrated beyond doubt.

In compliance with this rationale, it is specifically in the faults and the shortfalls of the classical models of production and management that one should seek the foundations of Asset Management. Moreover, that is how I was myself first confronted with the issue, as a young engineer hired as a maintenance manager of an infrastructure network.

If I choose to discuss industrial maintenance specifically and as a form of stepping stone for Asset Management, it is also because I appreciate that my situation as a former maintenance manager is far from singular. To this day, three decades after the first projects undertaken with a set of international colleagues, I still observe that a remarkable and continuous proportion of managers who commit to Asset Management or even to ISO 55001 Certification programs and approaches come from the field of maintenance on the worldwide scale.

Of the Necessity of Developing a Rigorous Asset Management Discipline

I often tell the tale of my first confrontation with failure: in that time, I was still working as a maintenance manager in my native Brazil, having recently graduated. I worked for a major public transports operator, and my position came with all that one could expect at the time in terms of tools and practices. Despite praiseworthy evolutions in the field of technological tools and practices, contemporary conventions and individual mindsets encountered in the maintenance sector have moreover remained almost unaltered since those days. We were stubborn and determined mechanics, and yet we paradoxically lacked any model of strategic vision for our task. Our workload was cadenced by a cycle of degradations and failures of the rolling stock and of the infrastructures over which we apparently had very little control, if not none at all.

Fatally, technical issues would arise. Machines would become inoperable and we would stubbornly strive to repair them, unquestioning of our absurd condition as "luxury spare parts exchanger," however qualified we may have been as engineers.

Still, restricting the action of such a set of engineers to interventions that irretrievably would be limited to the replacement of a defective mechanical part or equipment would not provide us with solutions fit to break this vicious cycle. Sisyphean tinkerers, we incurably repeated the same patterns, without stopping to wonder whether our function was the right one. In this regard, the arrival on the market of CMMS (Computerized Maintenance Management Systems) or Operational Reliability softwares have only comforted us in the idea that we were doing "a good job," by making the realization of the tasks we'd grown used to performing a whole lot easier.

My role was confined to repairing what could be repaired in the rolling stock, as it grew more ancient and the failure rate soared. This type of operation, inscribed in the fields of short-term and reactive activity, seemed to content my employers—after all, this was precisely why I'd been recruited. Yet, I gauged the contents of my warehouse and noted, privately, that we were operating in an irrational fashion. We reached our output goals, of course, and that was enough to satisfy my bosses—but even then it seemed clear to me that our management of assets and infrastructures derived from a "wait-and-see" mentality. Thus I felt a real sense of frustration, which would only deepen throughout the rest of my training, which was far from complete at the time.

With this in mind, I expressed my impressions to my former manager, only to find that he paid little attention to my reflections. A preventive maintenance–centered set-up, the likes of which I was beginning to envision, was, according to him, both costly and hazardous. Worse yet, when he would later come to adhere to the ideas I'd exposed, he would only do so out of self-consciousness and without resorting to a posture of technical objectivity in considering the issue and the models of realization that it should prop up.

There are millions of ticking time bombs in factories around the world. They might be quiet, but professionals in charge of assets know them well. At the very least, they suppose their existence. Some of them are sensitive to the ticking sound: they try to prevent the explosion but can only postpone it. Others pretend not to hear the ticking sound in the belief that they can do nothing to stop it. Thus, one way or another, sooner or later, bombs will blow up.

These bombs result from poor decision making in the field of industrial Asset Management. Such decisions are calibrated on the company's targets set by superior management, then travel through the company and down the ranks. When they reach the plant, their impact is considerable. Cuts will be requested successively in the budgets of maintenance, engineering, and in the modification plans projected on machines. An increase in production capacity will be imposed regardless of the mid- and long-term consequences. A decrease in the stock of spare parts will be decided with no consideration for the risk incurred by their absence. Machines will be physically operated well beyond their economic life cycle, and so on.

There are legitimate reasons behind these decisions: as globalization enhances competition, optimizing manufacturing and maintenance costs becomes vital.

As it was, business was going steady, which seemed to demonstrate beyond all doubt the soundness of the task I'd been assigned. For the first time, I was confronted with a duality that underlies to this day the reflections linked with the production and the management of industrial assets: that which opposes effectiveness and efficiency. In a somewhat Machiavellian fashion and under cover of either oversized investments or not-so-demanding stakeholders (often a combination of both of these factors), organizations have consistently repeated a pattern of putting their managers in a situation where they are more encouraged to "display service" than to deliver it in an optimal manner.

It was only a while later, as I was completing a post-graduate degree in France, that I finally managed to pinpoint and frame the precise purpose of my questions. A strikingly simple analogy dawned on me. The task I'd set for myself implied that I should behave, in my dealings with the machines, as a doctor would—but I'd never been handed the tools that would allow me to better understand their ills. I arranged a meeting with the dean of the nearest medical faculty, and asked for his guidance: "How does one make a young man into a doctor?" I enquired. The first year of medical faculty, he informed me, fundamentally relies on three complementary axes of study: primarily, anatomy (namely the physical composition of the human body); secondly, physiology (namely the functions that allow organs to interact between themselves); and finally, medical psychology, which qualifies the attitudes one should adopt when confronted with issues of illness and death. Biology and physiology, I mused, are rather self-evident; if one wishes to heal a body, it is logical that he should learn to know its particular workings. However, I realized at the same time that in my line of work, it was by no means required to be aware of what would constitute the "counterpart" of biology, namely the machines' *life cycles* and *reliability*. Hence, it was only after leaving my functions as a maintenance manager that I finally realized that the heart of the Asset Management matter lay in the disciplines of reliability and life cycle.

However, it was the third field of medical knowledge that affected me most profoundly: the psychology of death, which designates the idea that any organism will eventually collapse. Was that not the very notion that my peers so dearly lacked—these maintenance operators who stubbornly and beyond all reason strived to repair, heal, mend, and patch their machines, thus turning an eternal blind eye to their declining productivities and their ever-more-frequent failures?

As these thoughts were taking shape in my mind (in the early 1980s), NASA, together with the military complex, was developing *life cycle cost analysis* or LCC, a tool allowing one to assess the total cost of an asset throughout the entirety of its life cycle. This method was a much-needed refining of and complement to the *design-to-cost* approach, which was prevalent in those days. Indeed, it featured in its results the entire array of costs tied with an asset on the totality of its life cycle, as opposed to being restricted to the budgets allocated to the design phase. LCC made it possible to realize a long-term assessment of the assets based on anticipations over their technical and economic performance, thus facilitating the first step on the path towards a preventive-centered management policy. The ultra-liberal movement of that era, conceptually embodied by the theses of the Chicago school, favored studies focusing on the Net Actual Value and Internal Rate of Return to analyses that took into account the costs and expenditures tied with investment projects in the long run. As of that period, the Reagan administration took up the question of the privatization of municipality-owned transportation in the United States, which undoubtedly marked a decisive impulse in the democratization of life cycle costing, which has become a most common notion in contemporary economics.

That which derived from the synthesis of this innovative cost-analysis method and of the embryonic discipline of machine biology was the crystallization of a need to manage the asset on its integral life cycle; a need which evidently comes hand in hand with technical processes, but cannot be restricted to the latter since it is inevitable that inscribing such an approach in a long-term horizon will also induce an urge to create a specific management system. However, we now know that the extraction of value is only conceivable in a perspective that takes into account the entirety of an asset's life cycle and, therefore, its global economy and management. It therefore transpires that parallel to the emergence of a necessity (to move away from a deficient model derived from the theses of the Chicago school), the scientific community was equipping itself with a technical arsenal that made it possible to extract the value from assets on their entire life cycles—which is to say that we were gradually giving ourselves the means to undertake a journey specific to Asset Management.

We should furthermore observe that the blossoming of this new discipline was truly made possible only through the aggregation of competencies originating from diverse fields of knowledge. On one hand, we drew on the skills inherited from maintenance; on the other, we possessed a physical knowledge of the machines, derived from reliability-based studies, which had known a strong growth after the end of the Second World War; and finally, we had at our disposal a new perspective on economics, arising from the development of *life cycle costing*. These elements fueled and helped shape the new models of management that were arising at the time, and particularly those induced by the emergence of quality repositories.

It is therefore rather clear that the emergence of Asset Management was inscribed in a particular theoretical and economic context, and specifically one where new requirements were being set in almost every segment of the industrial world and its stakeholders. And indeed, the same types of practices were blossoming simultaneously in various regions of the industrial world; thus, one can observe that Dr. Penny Burns, one of the pioneering figures of our discipline, was then reaching very similar conclusions on behalf of the New Zealand government as early as 1984, in the context of her public infrastructures' revalorization project. One could therefore assert that Asset Management now relies on precise and rigorous concepts, and has been validated by empirical achievements. In order to shine a light on both of these central aspects, the following volume will be structured in a matrix organization, setting as its abscissa axis the different phases of the asset's life cycle and, as its ordinate axis, the theoretical and conceptual factors that impact these phases. This structure will simultaneously allow us to highlight the degree to which Asset Management is a transverse discipline, diffuse at every level of entrepreneurial organizations.

A Favorable Climate for the Emergence of a Practice

In order to understand the rising interest of the corporate and scientific communities towards the mechanisms of Asset Management, one must perceive how this evolution has occurred within a very specific economic and normative context. Indeed, the consequences of the 2008 financial meltdown have been profoundly disruptive. Between 2008 and 2013, several thousands of industrial plants have been forced to go out of business all across Europe. Furthermore, the part of the industrial fleet that has remained active throughout Europe is only running at an average 75% of its nominal output; while this can be partly justified by the weakening of demand in times of financial crisis, it is also clearly an effect of a severe lack of technologic renewal and innovation combined with an aging of industrial assets that is beginning to have a negative impact on organizational levels of productivity and competitiveness. Indeed, after a decade of economic crisis, we can now observe with certainty that the European productive apparatus has aged in disproportional excess relative to the number of years during which the influx of investments was insufficient.

How are we to envision this issue? Schematically, it can be established that the rarefaction of investments made it impossible for organizations to renew their asset fleets up to the expressed needs, therefore allowing for a continuous degradation of the assets already in place. We are therefore faced with a situation in which a rising proportion of assets are in their mature lives as opposed to their useful lives, in addition to the effects of obsolescence that threaten every industrial organization across the world. In more technical terms, one can observe the CAPEX/OPEX ratio has fallen (CAPEX: *capital expenditures*; OPEX: *operational expenditures*). Furthermore, it should be noted that in many organizations, this ratio tends to decrease for a number of years until a consistent break in production generates a massive influx of CAPEX concentrated over one or two years.

In spite of all this, assets remain the first post of capital immobilization and therefore for the Assets Base of these organizations. In the case of international organizations in which equipment and infrastructure networks make up most of their capital, as it can be seen for production, transportation, and energy distribution TNCs, one can observe as much as 150 billion euros of capital invested solely in assets. It would therefore be unwise to neglect the real weight of these assets in the mainstream economy. Furthermore, it is crucial to learn how to manage these assets with a systemic approach, even more so in a context in which asset renewals are rarefying; otherwise, one faces the risk of witnessing an uncontrolled, and therefore potentially dangerous, aging of the assets, which would threaten to endanger the business depending on these assets. Indeed, it is the nature of Asset Management to promote and implement a culture of objectified investment (regarding the parameters at play) in which the assets' life cycles are taken into account over their entire spans or over the course of their operators' liability periods.

The need to standardize and make available the accumulated knowledge in order to democratize the best practices in the field of Asset Management has therefore never been so strong. But to fulfill this objective, one must be armed both with a scientific arsenal and a strong operational experience, in order to make the true sense of these "best practices" obvious and to universalize the lore. Over the course of the last few years, institutions—primarily from the Anglo-Saxon world, but today on the global scale—have strived to create a conceptual and theoretical framework for Asset Management, through the publication of standards such as the BSI PAS 55. Regarding the ISO 55000/1/2 series, it should be noted that if ISO (whose standards act as a reference in 162 countries around the world) has deemed it useful to dispense an international standard for a discipline that had only been around for 20 years, it must mean that this knowledge brings an added value that has been urgently needed in the pre-Asset Management world.

Preliminary Notes Regarding the Essay's Layout

Asset Management may often appear to be a somewhat puzzling discipline, but it is also a noble one, in the sense that it sets out to implement a true pathway for the construction of a structure of decision-making.

As we're well aware, any decision taken within an organization, regardless of its success, is always dependent on the action of a leader, a human agent prone to making valid, disputable, or misled decisions.

Therefore, we set the purpose of this work in the perspective of the construction of decisions, without denying in any way the vital part played by the decision-makers themselves, who remain fallible human actors and stakeholders in the Asset Management process.

Why should this principle be regarded as a major guideline for this essay? The explanation is rather simple: beyond the transdisciplinary and multidimensional vitality of Asset Management, it has become clear that the two initial chapters of the ISO 55001 standard place our discipline under the two auspices of understanding the specific context inherent to each organization and of the notion of leadership, namely the ways in which roles and responsibilities are dispatched and deployed within said organizations. We must therefore respond to this observed impact of the leadership's performance in the mobilization of the organization's human energies in the drive to generate the profits inherent to Asset Management.

Thus, this essay was produced with these two crucial segments in mind: leadership on the one hand, and proper usage of technical and technological processes on the other.

Value extraction and long-term success are the only indisputable signs of a valid discernment in the context of Asset Management. But they would remain incomplete if efficient decisions did not aim at satisfying the conditions for a proper supervision of their implementation; in other words, not only is it vital to make the proper choices, but one should also be careful to ensure their "executability" without losing sight of the attached economic dimension it entails.

To deal with these seemingly disparate issues, we've chosen to organize the essay around four segments. The matrix provided below serves as the departure point for our reflection, which strives to provide the reader with a conducted tour through the various sequences of an asset's life cycle in the order of their occurrence, from design to dismantling.

This choice is easily justified: the maturity of the professionals involved in Asset Management is on a constant rise, and since the launch of the ISO 55000 standards we've witnessed a considerable refinement in the questions asked by "newcomers" to our field. The notion of "life cycle" is among the main points of interest, and it appears as a major conceptual break for professionals who wish to distinguish thoroughly Asset Management from the other disciplines that aim to support industrial performance (the most obvious example being that of maintenance methods and activities).

An alternative and more personal justification for the elected layout of the essay resides in the fact that a decade ago, I published another work that sought to "give a voice" to the machines and assets, in a deliberate attempt to steer clear of methodical dogmatism. At the time, there were no existing international standards for Asset Management, and it seemed that to produce an array of conceptual answers to questions that were still in gestation would not have been a very rigorous approach. In today's context, however, and in order to avoid the bad habit of methodological "guruization," it was opportune and timely to produce an orderly and detailed account of the entire life cycle of assets; hence our decision to follow the trail of the full cycle of the assets' existences, from their "project phases" to their "end of lives," and spanning their preoperational and operational lives.

Every one of these chapters was designed in such a way as to shine a critical light on the common practices in the specific segments of the assets' life cycles: indeed, even though these practices are often of a superior level of quality, they fail to construct a holistic rationale bent on the perspective of long-term value extraction. Therefore, we've chosen to proceed to a number of historical reviews, in order to document the evolution of technical processes and the various conceptual breakthroughs in the field of Asset Management, while sticking to our wish to "demystify" the overly segmented and disparate function of typical organizations. This will open the way for the development of positive perspectives of alignment, foresight, and refinement of the decision-making processes.

We strongly wish to impact positively the readers' reflections regarding any commentary or criticism of historical facts that could be understood as an expression of disparagement or belittling. Our discourse is not one of "creative destruction," in the Schumpeterian sense; on the contrary, we regard this process as the natural spiral of a discipline in full expansion. Our posture is therefore one that seeks to describe a positive paradigm change, rather than one of unconstructive criticism.

Despites its frequent references to the ISO 55000 series of standards, this essay is not a user guide for these standards, complete with the full set of structural guidelines that would facilitate an industrial agent's claim for eligibility for this certification. Our focus is set on the decision-making process in Asset Management, and therefore on the necessary quality of discernment required from every industrial agent in order to manage the relevant elements that must be combined in the face of every question, issue, dilemma, or crisis relative to the management of assets.

It is quite understandable that the reader may sometimes be surprised by the apparent "redundancy" of elements or themes in various chapters of the essay. My hope is that the reader will not hold it against me; indeed, the diversity of the disciplines that have a part in Asset Management and the complexity of their intertwinement demand that these elements be repeated rather than presented in an incomplete fashion in order to ensure that their explanation is sufficiently clear and thorough. This is the reason for the thematic diversity presented in the lines of the matrix provided below, whose elements were treated in a non-linear manner according to the specific stakes of each segment of the essay.

Preface

Celso de Azevedo has a reputation in the field of Asset Management that always precedes him. I had the privilege of meeting Celso for the first time at the inaugural meeting of ISO/PC251 (Asset Management) in Melbourne, Australia in 2010. When I met him and engaged with him, I quickly learned that everything about his reputation was true. A man with a heritage spanning Brazil and France, bringing the passion we expect from those cultures to all things Asset Management in a larger-than-life way, but merged with a vast experience and ability to analyze and learn.

In our ongoing work on Asset Management in ISO/TC251 (we changed from a Project Committee to a Technical Committee with a wider scope in 2015) our focus is on helping organizations deliver "value." For many people, the word "value" is interpreted as a financial value only, but we live in a complex world with complex stakeholder groups, each with competing demands and competing definitions of what "value" means to them. That value may be financial but it may also be safety, or environmental or ambience or reliability or a myriad of other tangible and intangible benefits depending upon the respective viewpoints of those stakeholder groups. Pleasingly, the core of this book is "value" and how it is overlaid across the asset life cycles and how it relates to costs.

This is a nice approach because in our organizational worlds, we may only have a single window through which to view one part of the asset life cycle, so for all readers there will be at least one section of the book with which they are familiar. The remainder of the sections are areas with which they should be familiar, and this is the Asset Management paradigm. Namely, we must have an understanding of all life cycle phases *and* all organizational functions even when our own function and window on a specific life cycle phase may be dominating our daily thoughts.

If this is not complex enough, Asset Management is a young discipline with many evangelists claiming to know the one true way! This leaves any author with a dilemma on how to balance the need to communicate a new discipline in an authoritative way to those who want to learn, while at the same time, hypothesizing and exploring how the discipline will grow. All of this needs to be achieved without appearing to shout from an ivory tower. No simple task at all!

Celso has written this book in a way that enables us to explore Asset Management thinking throughout the life cycle of the assets but with a focus on value (both tangible and intangible), recognizing the importance of cost and how we express intangible value through the measurement of monetized risk. He does not claim to have that singular truth of how to solve all Asset Management problems but instead presents his perspective on those problems through his extensive experience both from infrastructure and industrial scenarios around the globe. He presents theories for solutions, not in a mandatory way, but more in the approach of the traditional scientific method. By this I mean that this book is an invitation to challenge and test his theory through our own experience and application. This addresses, I believe, the problem of the evangelist author and his singular truth.

What of the problem of writing about complex organizations and complex asset life cycles? Inevitably in writing a sequential book it is required that the text be divided into chapters. Some readers will be drawn to individual sections to learn or even to challenge. Others will read the whole book from front to back. I am sure Celso would invite us to learn and become expert in each of the areas explored but having known him for many years, I am also sure that he would advocate total adoption of all of the thinking to some extent. For me, this reflects the complex nature of Asset Management, where we need a broad view of context and the ability to apply analytical approaches in great detail into that broad context.

In another potential life where Celso de Azevedo was perhaps Dutch and not Brazilian/French, he may have termed it Total Asset Management like the great Dutch soccer teams of the 1970s. In Total Football, a player who moves out of his position is replaced by another from his team, thus retaining the team's intended organizational purpose and approach. I feel there is much of this book that reflects this thinking and I hope that you enjoy it, learn from it, test it, and challenge it to deliver greater value for your organization's stakeholders.

Rhys Davies
Chair of ISO/TC251—Asset Management

The Structural Approach to the Book

The Matrix of Methodological Elements Associated with the Life Cycle Segments of Assets

Elements of Asset Capability I've taken the liberty of color-coding the elements of Asset Management as three different segments: light gray for pre-op, dark gray for operational, and white for the end-of-life, wh could serve as "chapters" in which to divide up the book.				
	Design and Conception	**Evaluation**	**Specification**	**Procurement**
Value to • Stakeholders • Corporate • Employees • Public • Etc.	The finality and specific nature of the assets must be the driving element of the value-to-extract	Taking into account more that strictly the economico-financiary technics in the definition of values	Need to the interests of stakeholders/ alignment of value	Taking into account of values far beyond purchase price
Constituents of Asset Capability				
People • Engagement/ commitment • Behaviors and values • Competency • Etc.		Anticipation of the life-cycle	Taking into account future needs in terms of competencies linked to the value-to-extract	Raise awareness regarding life-cycle timely for commitment
Finance • Accounting • Budgeting • GAAP tools • Technology		Life-cycle analysis vs. design-to-cost/ monetization of the risk factor (including technology)	Alignment between asset register and asset costing	Life-cycle cost process

Installation	**Commission-ing**	**Operational**	**Maintenance and Modification**	**Replacement vs. Disposal**
	Optimizing profit from this run-in period by also verifying chances and real values to be extracted	Having an active consciousness of the fact that everyday we can optimize the extraction of value	Realizing that "parts replacement" and "value extraction" are not necessarily compatible	The values of an organization are so many drives for end-of-life arbitration
Impact of BIM on competencies profile	The worst and best moment for understanding competencies required	Adequation between competencies and evolution of the needs, between level of commitment and values alignment	"Loving care" behavior and emotional attachment	"Loving care"/ rise of a need for an aging management structure and roles
Alignment between asset costing/as set register	Budgetary differentiation between commissionning and operational phase	How asset costing feeds the decision-making process	How asset costing feeds the decision making process	How depreciation rules and provision policies can impact on the timely replacement

	Design and Conception	Evaluation	Specification	Procurement
Management Systems and Processes • Decisions • Technology • Types of Processes • PDCA • Etc.		Alignment of corporate objectives		
Asset Capabilities		The historical decline of life-cycle costing techniques The importance of FEED approaches which take into account risks and costs in project phase	Success and failure factors of the insertion of the LCC in specification Demystifying untenable counter-arguments	The clear and obvious need for LCC in buying processes
Risk and Decision-Making • Stakeholders • Corporate • Asset • Deterministic vs. Probabilistic vs. Legal • Technology		Demystification about the interdependency between different factors (stakeholders/corporate/regulation)	Impact of the error reproduction on new asset portfolios: "avoiding copy-paste projects"	Status about the trade-off between corporate desires and technological prices

Installation	Commissioning	Operational	Maintenance and Modification	Replacement vs. Disposal
Milestone for key elements in the AMS (configuration management)	Alignment of corporate strategies		Theory of signals: Necessity of an aging manager to assess the asset's end-of-life status/ineffectiveness of practices such as "modification redesign"	Theory of signals: Necessity of an aging manager to assess the asset's end-of-life status
Reliability, redundancies	To comprehend this phase as a hiatus from the guiding trade-offs	The importance of operators in the examining of trade-offs The necessary distinction between assets in service life and assets in mature life	The necessary and daily usage of trade-offs and other optimization tools	Mastering the mature life and the trade-offs specific to the asset's end-of-life
Status about the trade-off between corporate desires and construction prices	Avoid allowing the difficulties on the commissioning phase become the drive for the life-cycle	Maintain alignment with prior objectives and commitments	How to make "luxury piece-replacements operators" evolve into real business engineers (linked with "People")	Curve of total impact/accepting the asset's end-of-life and the interest of timely replacement

	Design and Conception	Evaluation	Specification	Procurement
Trade-Offs and AM Objectives Desired trade of: • Cost • Risk • Performance		Trade-offs CAPEX/OPEX Construction of technological scenarios based on FEED simulations (RAMS studies and costs against stakeholders values)	Trade-offs CAPEX/OPEX Definition and selection of scenarios based on the company's short-term and long-term values	Trade-offs CAPEX/OPEX Definition and selection of scenarios based on the company's short-term and long-term values
Metrics and Measurements • AM objectives (levels) • Use of KPIs • Etc.		Translation of the values into coherent METRICS ahead of the asset project	How to make the values-to-extract self-explicit in the case of functional and material specifications	The convergence of operational purchasing measures (tied with specifications) with AMS rules linked with the extraction of value

Installation	Commissioning	Operational	Maintenance and Modification	Replacement vs. Disposal
Reliability, capabilities (measurement of the maintainability, etc.)	Confrontation between levels of reliability (inherent vs. operational vs. real) Future contractual specifications in asset management	Daily usage of trade-off techniques and the construction of decision-making scenarios	Exploitation and maintenance, "same battle" Daily usage of trade-off techniques and the construction of decision-making scenarios	Huge deficiencies and need for a mastering of the concepts and techniques of arbitration regarding assets in end-of-life phase
Metrics eminently tied with project monitoring	Be aware of the difference between KPI measures in this phase and those which come next	Learning to create indicators and metrics specifically tied with the values to extract in each organization	Learning to create indicators and metrics specifically tied with the values to extract in each organization	Create handy milestones of useful and mature life so as to optimize the extraction of value in end-of-life phase

Acknowledgments

The field of advising good practices is deeply indebted to the innovations that have sprung from reflections around the problems to which existing solutions were so far unsuited. Obviously, Asset Management is no exception. In all logic, I would like to thank the life cycle of assets in the first place, and the attached necessity to think in a systematic manner in order to improve our performances. Without it, Asset Management would probably not exist.

Similarly, I must thank the contradictions that we are immerged in daily. Without them, there would be no need to seek out new equilibriums to base our steps and our decisions on. We would like to be able to tread these new itineraries with certitude. Yet these do not really exist. I therefore thank my work in Asset Management for making me aware of these uncertainties, and for the humility it has taught me: it is a necessary quality for any human and scientific progress.

Taken alone, these factors were not sufficient to incite me to write this book. The industrial field and its infrastructures, with its pressing demands and inevitable urgency, challenge our ideas, our outlook,and our behavior in order to comply with the expectations of results, or at least their necessity.

At the core of all these elements are the people closest to me. I am fully aware of the extraordinary effort it was to consent to help me with this task, and I am grateful for it.

To Edmea Adell, my partner throughout this journey of learning, implementing, and passing on the teachings of Asset Management. I owe her infinite thanks for the quality of her advice, her persistent support of my work, and her unfailing dedication to the promotion of Asset Management in France and around the world.

To the teams of Assetsman, longstanding colleagues, and excellent professionals in France and in Brazil, with whom I share a task that requires high levels of knowledge, rigor, and dedication. Handling the tools of Asset Management for our clients is the focal point of our daily lives and of our work in common.

To Benjamin Carde and Emile Poivet, whose work on language has made this book fluid enough to satisfy the expectations of any type of reader.

Finally, a special inscription for my sons, Gabriel and Julien, who have grown to become dedicated professionals in their own fields, and who have already assimilated the difference between "looking after assets daily" and "the Asset Management in the long run."

Asset Management Insights

Phases, Practices, and Values

PART I

THE PROJECT PHASE OF INDUSTRIAL ASSETS

Unlike Industrial Maintenance, Asset Management is by no means limited to the proper conversation about assets throughout their operational lives. Indeed, when we state that the finality of our discipline is to accompany and optimize the extraction of value across the totality of the assets' life cycles, we imply that the prerogatives of the Asset Manager cover fields of temporality and management more diffuse and more spread out than those devolved to the maintenance operator.

Thus, the preoperational phase is primordial in the context of an optimal management of assets, because it allows for the implementation of fundamental principles in terms of objectives' alignment and risk monetization. Furthermore, a set of recently developed techniques and processes allows us to evaluate with an appreciable precision the impact of the decisions taken by consulting offices in the very first phases of the design and conception of industrial assets.

CHAPTER 1

Demystifying the Project Phase

1.1 The Rise of a Praxis

In the classical industrial scheme, the preoperational process essentially depends on two factors in its effort to acquire or design an asset. On the one hand, it depends on the efficiency of the tasked design office; on the other hand, the budgetary constraints imposed by the industrial leadership of the office. It should be noted that this approach is inherently dangerous, not least because it implies short-term savings, which will indubitably have a negative impact on the reliability and the efficiency of assets throughout their life cycles—and which could well result in a heavy and undesirable financial expense later on. Furthermore, the compartmentalized approach of the industrial organization chart, which opposes engineers and financiers, strikes us today as inefficient and anachronistic, whatever we'd like to think of the issue. Indeed, the alignment of objectives, which allows for the maximum extraction of value, cannot occur in a context of separation of powers within an organization.

Research on the means by which one can improve the preoperational life of assets was not born from Asset Management: indeed, as early as the end of the Second World War, engineer Lawrence Delos Miles had begun to reflect on the most adequate methods of design while working for General Electric, and he consequently came up with an approach nowadays known as *value analysis*. This innovative process prefigures, albeit at an embryonic level, the approach championed by Asset Managers in the sense that it connects the notion of economic potential to a fundamental pragmatism that aims to limit costs in a rational manner.

It is worth remembering that in 1973, the first crash of the oil market brought about a meteoric rise of production costs at the organizational scale. Savings became the primary and central constraint in the engineering sphere—designers were thus requested to take into account stricter and stricter budgetary constraints, which allowed for value analysis, in the form of *design-to-cost*, to take a durable foothold in the industrial culture of conception. It would be unwise on our part to make a value judgement on this trend, seeing as it was above all a response to a new economic context where budgets were appearing as the primordial criteria of reflections on conception. Therefore, design-to-cost was tied to the issue of competitiveness, or even of economic survival.

Nonetheless, it is clear that if we consider it from the perspective of an Asset Manager, design-to-cost is inherently a flawed model. Hence, it does not take into account life cycle costing or value extraction. Indeed, this approach of design takes its base solely on CAPEX, and therefore on short-term investments, without generating any reflection on OPEX (the costs tied with the operational phase) or on the life cycle. We could therefore postulate that despite its innovative touch, design-to-cost is diametrically opposed to the practices and rationales inherent to the logics of Asset Management.

This argument is in fact tied to a very contemporary problem, in the sense that the rationale of savings (which has derived from the budgetary constraints of the 1970s) has consistently influenced corporate culture to this day.

Thus, organizations have not been able to break free from a certain approach that places the limitation of costs as a finality *in itself*, a fact that can largely be attributed to a lack of sensitization on behalf of management experts—in this case, of Asset Management experts. Yet, we know that the rationalization of costs depends primarily on an analysis of the predicted reliability of the assets and on an appraised anticipation of the risks tied to these assets—a problem that design-to-cost inherently neglects. It is therefore up to the professionals and experts of Asset Management to democratize a new techno-economic culture and to shine a light on the limitations of an anachronistic model.

All too often, industrial and infrastructural managers base their decision-making on the sacrosanct financial indicators derived from the theories promoted in major business schools worldwide, while feigning to ignore that a capital-intensive installation will always be faced with crucial equipment reaching its end-of-life and whose renewal is therefore vital to the proper function of business. If they are seldom viewed as a priority according to these indicators, it is, however, clear that these end-of-life assets will have to be replaced, regardless of their ROI (Return on Investment) or of their IPR (Internal Profitability Rate), and despite the organization's *cash flow* situation. In fact, they will have to be replaced even at the cost of postponing other, more profitable investment options.

We can't overstate the fact that in the Asset Management perspective, the value of an asset is to be extracted over the course of its life cycle. This is because the reliability of equipment (as opposed to purely economic indicators) is not recycled according to fiscal years (or capitalization periods) but according to the "biological" life cycle of machines, which cannot be reproduced[1] to fit the comparison of NPV calculations when the economic frames of concurring projects differ.

A few lines earlier, we deplored the lack of an "appraised anticipation of risks" in the context of design-to-cost. One should always be careful to demonstrate a methodical conceptual rigorousness when dealing with such a vast topic. It would be misguided to assert that risks are not taken into account in a serious engineering or industrial system. It is quite the contrary, as they have always been at the core of the theoretical priorities of design managers; however, risk monetization, which refers to its appraised quantification, is a phenomenon that has only emerged with the rise of Asset Management in the 1990s.

Until then, the acknowledgement of risk, dictated by a minimal or nonexistent *corporate* and *preoperational* alignment, was not a qualitative indication—and was bound to remain, therefore, both limited and imprecise. This was not the result of bad will on the part of the designers, but rather of a lack of technical means, especially in terms of calculation—a strong hindrance to an efficient strategic program. Indeed, U.S. Military Standard 1629A (the FMECA standard that defines risk analysis and the criticality of engineering systems) was only published in 1967. As it happens, every notion impacting on design and the ponderation of values (including risk) made its way into the scientific discourse around the same era and slowly grew in precision and clarity. The place of risk within the process of analyzing the value of an asset in its preoperational phase was therefore very different at the time than today; it was perceived as a "ponderation criteria" but its monetary translation was very marginal in design offices.

In order to give more substance to this outline of a reflection on risk, one needs to be equipped with a technical arsenal that allows for a quantitative assessment of risk itself. This was the main input of life cycle costing, or LCC, a method developed by NASA and MIT researchers for the benefit of the U.S. military in the late 1970s. It should be noted that the great economic reforms pushed by Margaret Thatcherand Ronald Reagan are very much tied to these initiatives. For the first time, the entire life cycles of assets were taken into account from the earliest stages of the preoperational phase. However, the transition between these two opposite poles (design-to-cost and life cycle costing) was not a simple one, and it would take many years for a rigorous economic translation of technical risks to see the day.

This historical perspective on the evolution of the way *value* has been perceived is very telling. It shows that the notion of quantified anticipation of an asset's value as well as reflections on the most appropriate means by which to extract its optimal value are very recent fields of thought that have yet to become regarded as utmost priorities within design and engineering offices.

1.2 Designing: A "Distinct" Activity

Throughout my career, I've been given the opportunity to observe the various phases of the assets' life cycles in numerous contexts, and from distinct perspectives inherent to the specific affectations I'd been given. It must be noted that the preoperational phase of industrial assets is all too often regarded as an activity "within itself" in the industrial process. Thus, it is broadly believed that it is "distinct" from other phases, which implies that it has its own "end," in the same way that the processes tied to this phase of design or procurement themselves have a beginning, an evolution, and an end. Intuitively, this can appear as a reasonable belief. However, it is in fact a valid indication as to the gap that yet prevails between the traditional industrial culture and the perspective offered by Asset Management culture. Indeed, regarding the preoperational process as a "distinct" phase marks a clear opposition to the very notion of an asset's global life cycle.

Ever since prehistoric times, the social man has striven to better his production of tools, then of machines—in short, of assets. But up to the Industrial Revolution, and, to some extent, even in our modern times, assets that were to be operated have been regarded as ends within themselves, thus neglecting the optimization of value whose realization they would permit. Reflections of life cycles have not been useful up to now, because the notion had not been proven to embody a true necessity. Historically, designers have been allowed to create in total freedom from external constraints, since what was expected from them was to design a tool responding to a precise need, and whose function was immediate.

Until the introduction of CAPEX/OPEX trade-offs (which establish a relation between procurement investments and maintenance costs in the production phase), costs had been somewhat absent from this historical scheme. Since the task that had been set consisted solely of responding to a functional imperative, it seemed unnecessary to take into account risks of degradation and aging of the assets and machines. Thus, I've been invited a countless number of times to take part in broadly chance-based projections of operational and maintenance costs, sometimes as early on as in the context of a call for tender where all too often the practice, even today, is to appear as the concurrent able to display the smallest "ignorance coefficient" in regard to the operational costs of the life cycle. We can take pride in having participated, throughout the last three decades, in the implementation of a culture which takes operational reliability and life cycle costing much more seriously; however, this culture has yet to impose itself as a true industrial standard.

To this day, one can deplore a lack of anticipation regarding the life cycles of different assets. Even in the case of organizations that have begun to open up to an Asset Management line of thought, the profile of professionals sought after for this type of mission is radically distinct from that expected for other missions. This can induce a lack of coordination, and therefore of *alignment*, in corporate perspectives. Once more, we can observe that organizational segmentation is harmful to the optimal efficiency of decisions tied to the assets. The reality is therefore that of a complex mix of needs: that of the implementation of a more "long-term focused" vision, of a transformation of traditional leadership, and of a more holistic management for the different segments of the assets' life cycles.

Let us stop to consider the role attributed to design offices in the organizational chart organigram: these offices are responsible for the assets' conception, design, and specification. Thus, they are exclusively made up of engineers, trained and habilitated to consider the asset in terms of its sizing and functionality, but not specifically in terms of its life cycle.

By no means do we intend to minimize the usefulness of the engineers' work, which is more often than not very well executed and remains entirely necessary; our intention is rather to bring forth a redefinition of the value that this engineering work generates. It is obvious that engineers produce a certain value, were it only of a commercial nature. But in order for us to be able to talk about "value" in the sense of "value extraction," inherent to the best practices of Asset Management, the designer-engineers themselves should be taught and trained on the fundamental concepts of Asset Management; once more, the absence of a proper coordination of the different segments that make up the traditional organization is an obstacle to the optimal realization of value from the assets.

This "coordination" that we've brought up allows for an optimization of the alignment between corporate objectives (expressed by the organization's decision makers) and the work of the engineers. It is a crucial factor in the creation of a high-value asset, and one whose importance can be observed from the very first moments of the preoperational phase. In order to achieve an optimal creation of value, the action of the engineers must be similarly aligned to that of other branches of the organizational chart. Indeed, in the notion of *Line of Sight*, introduced in the referential document BSI PAS 55, it is stated that it is crucial to take into account the expectations and interests of stakeholders and to ensure that these are aligned with operational actions from the earliest stages of the preoperational phase as they are the more intricately tied with the assets' proper financial and physical health. They are, in the end, the most prevalent factor in the achievement of the required performance.

1.3 Decision-Making in the Preoperational Phase: Atony Comes at a Cost

In the overwhelming majority of organizations, we can observe that a vast number of industrial projects are bought or designed on behalf of preexisting rationales and by the means of a reproduction of anterior projects. Industrial decision makers often try to justify these practices through commercial agreements (a typical example would be that of turnkey projects) or through political and economic reasons. Organizations benefit from this approach on the short term, by making economies of scale and by radically reducing the temporal and budgetary resources mobilized for conception. However, this "recooked" approach to the design phase is in stark opposition to the inherent principles of Asset Management and to every notion brought forth by studies on *life cycle costing*.

Let us consider, in order to exemplify this line of thought, the case of industries that rely heavily on continuous processing (such as metallurgy, chemicals, etc.) or on massive amounts of fixed capitals (such as automobile manufacturing or large-scale infrastructures, e.g., water processing). In four out of five of these cases, the mechanical installations (for example, a line of laminators) are purchased "key in hand." It would be disproportionate to expect from organizations that they reinvent their entire field of activities for every new project that they implement. However, the present rationale, which takes its roots on the one side in the faith placed in a certain economic illusion on the part of the decision makers and, on the other, in an analysis betrayed by a profound misunderstanding of the machines' life cycles, is strongly problematic. This is because those who abide it remain, despite their best intentions, unaware of the changes that occur every day in the spheres of economics, finance, and technology, as well as of the needs and interests of stakeholders.

To indefinitely reconduct such projects is a symptom of a refusal to recognize that the world is in permanent mutation and that it is both *responsible* and *proactive* to try and adapt to its evolution. On the contrary, acting as though "one could bathe twice in the same river" appears as a posture of pure negligence.

The teams of engineers that operate in the design offices of these heavy industries are well aware of the necessity of associating metrics of performance to every project being developed. But those very metrics have been designed in a frame that is constrained to the task of conception, or the "project" phase. Thus, the measurement units that are used to qualify projects are unfit to assess their performance in their mature lives, and are therefore inefficient or imprecise at best.

1.4 An Opportunity for Improvement in Asset Design

Asset Management strongly supports the creative freedom of designers and conception agents since it is an undeniable vector of the rise of a new holistic perspective, which takes into account the assets' complete life cycles. It is therefore essential to encourage the innovative work of engineers. In order to do so, a number of techniques have been developed thanks to the input of *life cycle costing* and *trade-offs* studies. Not only are these techniques now broadly available, but their efficiency has been demonstrated empirically.

Why, then, do managers not strive to more efficiently align objectives as early as the preoperational phase? Why do they not attempt to master practices such as risk monetization? Indisputably, a number of them have already begun to do so. However, they only make up a marginal portion of industries so far and, in most of these cases, the methods employed remain rudimentary. Yet the shortfalls which derive from this posture are nothing short of phenomenal. The growing democratization of the availability of applications and software relying on very powerful algorithms should contribute to the evolution of these "modes de travail," which are still widespread.

It has never been so easy for an organization to achieve the alignment of its objectives and the monetization of its risk factor. The tools and skills required for these tasks are now universally available. The fantasy, deeply rooted within the industrial spheres and which depicts Asset Management as a field of experts estranged from economic realities, must necessarily be revoked. It is crucial that the agents of Asset Management succeed in sensitizing organizations to the added value that would come out of a proper management of their industrial assets, and to the necessity of implementing such a management.

Today, we are capable of drawing out and quantifying the entire set of economic consequences resulting from a good or a bad CAPEX. Why should we not do so? We are also able, when choosing or designing an asset, to anticipate the technical and economic consequences in both qualitative and quantitative forms (through previsions tied with OPEX and risk). Once more, why should we not do so? By what logic can a design office manager consider that it would be wiser to refrain from implicating an Asset Manager? In this regard, the stellar rise of the *Building Information Modelization* (or BIM)—an approach very similar to

Asset Management practices, and which has quickly been adopted in the sphere of construction—is exemplary. Industries cannot allow themselves to pass on such opportunities any longer; they must, consequently, equip themselves with means of value anticipation. The future will undoubtedly confirm my intuition that Asset Management will be the spirit (and BIM, the skeleton) of the infrastructures of tomorrow.

Through these examples, we have attempted to show that organizations must become aware of the importance of a rational management of their assets from the earliest stages of their life cycles. We will now draft out a few of the methods, inherent to Asset Management, which allow industries to establish an optimization of the creation of value in the preoperational phase.

At the core of this notion are the ideas of Asset Breakdown and Asset Register. The former can be defined as a tree-structured model, which presents the functional and material components of the asset that is to be acquired or produced; the latter is an accounting method, which consists of taking an exhaustive inventory of an organization's assets fleet. Therefore, it is clear that a thorough practice of the Asset Breakdown allows for a better definition of the parameters and the contents of an organization's Asset Register.

Today, the Asset Register has become a *sine qua non* requirement for an operational asset system to be in compliance with the current standards; how practical, for contemporary industrials, that equipment is now systematically delivered with its Asset Register (which entails the complete listing of an organization's assets categorized in at least three sub-classes: Asset Portfolios, Asset Systems, and Individual Assets). One should keep in mind that it has not always been so, much to the detriment of operators. Additionally, the massive democratization of ERPs—these tentacular informatic programs that link every organizational function—has also introduced new protocols of intra-tree structure relationships between assets, often comprised of nine levels (familial hierarchy between the assets of the Asset Register). In some alternative cases, ERP editors have imposed the use of the standardized protocol defined by the ISO 27000 (Security of Information Systems), which also presents the data imputable to the assets on different hierarchic levels. We should therefore rejoice over the growing influence of Asset Management inputs on the very structure of the industrial world, since in this case we've truly witnessed, throughout the last few years, a process of clarification and standardization of Asset Register practices, to the extent that they can now be regarded as truly reliable.

This first phase of an asset's life cycle, which covers its breakdown and its inclusion in the Asset Register, is much simpler nowadays than it had been in the past. Thus, considering how useful it can prove to be in the long run, it is now unjustifiable for an industrial manager to accept that this phase be neglected. Indeed, this process constitutes a stellar opportunity to effectively connect the preoperational phase with the remainder of the assets' life cycles.

If an industry succeeds in making it so that tomorrow its Asset Register is more proactive and more effective in taking into account the entirety of the assets' life cycles, then this industry will be ripe for establishing what is known as a one-to-one correspondence between the Asset Register and the Asset Costing. This step is a crucial one in the path towards implementing a strategic alignment, which, in turn, may give way to techno-economic alignment.[2] Indeed, ISO 55001 requirements clearly demonstrate that operators have much to gain by having one-to-one correspondences between physical assets and their attached economic and accounting existences. In a truly functional and optimal alignment system, one must be able to identify physical codes that clearly correspond to the equivalent accounting codes. If this is the case, one can speak of a "paired" system.

Over time, the engineers who make up design offices have been able to develop very exhaustive approaches to the preoperational phase, in order to address every important question raised throughout the design, specification, and procurement stages. More recently, in parallel to the development of reliability anticipation techniques and to the emergence of systems that allow for a modelized simulation of technical and operational performance, the engineering world has witnessed the rise of FEED (Front-End Engineering Design) approaches (see Figure 1.1).

The FEED approach relies on a modelized simulation that primarily takes the form of *Reliability Block Diagrams* (RBD). Thus, Asset Management generates models that will allow engineers to visualize very early on in the process of architectural and technological arbitrages whether the assigned objectives, in terms of productivity and operational availability, are optimal. Nowadays, one can draft out scenarios of economic consequences—and not only of technical consequences—from these different simulations and project options. The qualitative input obtained from this type of approach is furthermore immense.

Obviously, even if one relies on such modellings, a number of uncertainties remain; however, we now have the tools to scan these scenarios and assess their levels of sturdiness (in mathematical terms). Thus, one can evaluate these scenarios both qualitatively and quantitatively (through the analyses of "best," "worst," and "basic" cases) in order to determine in an informed manner the ideal course of action.

These new methods endow us with a true choice panel in the sense that one can now elect to reproduce a design or to innovate completely, or even to freely reimagine the solutions that engineering may procure. One can only hope that these solutions will aim towards more value extraction in the long term rather than stopping at the oft-imaginary or self-imposed constraints that until now, impeded the horizons of potential scenarios.

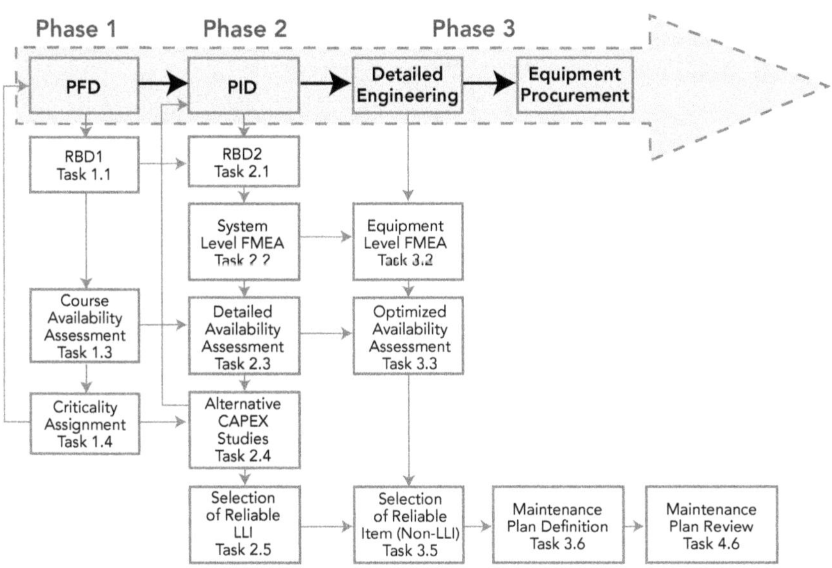

PFD—Process Flow Diagrams
PID—Piping and Instruments Diagrams

FIGURE 1.1 FEED Process—Front End Engineering and Design

CHAPTER 2

Reimagining the Procurement Function

In light of the remarkable weight of procurement operations within industrial budgets, it is quite clear that the procurement function plays a central role in organizational charts. Nonetheless, because of budgetary constraints or of imperatives tied with shareholder demands, this function is all too often reduced to a simplistic search for the minimal level of investment. This implies that more often than not, we observe that acquisitions don't follow a rationale of clear and anticipated amelioration of value-generating processes for the organization, but rather derive from short-term calculations; this creates a situation where procurement appears primarily as a factor that aims to conserve shareholders' interests through a reduction of CAPEX.

It is therefore quite clear that the generalized practice of the procurement function as it can be observed in recent years is in stark contradiction with the policies encouraged by the adepts of industrial Asset Management. We will thus try to go over the issues and the perspectives that derive from the modus operandi of procurement managers in order to bring forth an innovative and rational perception of what the procurement phase should become in order to comply with the requirements of a true "Asset Management" line of sight.

What is it that we find so reprehensible in the historical posture of procurement agents? And why do we oppose their practice so stubbornly? The answer is truly quite simple, but it shines a light on some of the cultural and internal rationales that have a lasting impact on corporate decision-making.

2.1 The Fixation on Cost

It is crucial to consider the very specific situation of the procurement phase in the overall life cycle of industrial assets. Indeed, this segment is regarded as being neither under the responsibility of engineers, nor under that of operators. It is a function whose limits are very well defined and almost partitioned—which comes with a fair share of problems. De facto, procurement has always operated in an isolated manner relative to the rest of the life cycle. Yet, the theory of constraints demonstrates that the optimum of the sum is not equal to the sum of the optimums; this means, in this context, that it does not suffice to adequately administer the procurement segment in an isolated manner in order to improve the global performance of the organization. For the enthusiasts of Asset Management, this is a systematic problem whenever one seeks to realize a function's optimum without laying a simultaneous view on the entire life cycle.

No one is exempt from such pitfalls: in my own experience as a manager, I've conducted procurement deals without giving preliminary thought to their long-term impacts. I've done so without a hint of embarrassment, as it seemed at the time that I was doing what I was meant to do. Indeed, what is it that is expected from a buyer? Based on undeniably variable technical criteria, his role is to identify and select the correct service or the right product, in compliance with current specifications, and always at the lesser cost. Regardless, it seems obvious that the role of the buyer should be broader and more proactive: his function should come close to that of an internal consultant, or in other words to that of an agent capable of assessing the needs and wishes of every segment of the organization in order to establish a coherent line of action and to keep expenses under control. And de facto, there have been many procurement managers who have exercised their job in this fashion; however, over time, the increasing weight of economic pressures has reduced the role of buyers to a mere speculation on prices.

This rationale has reached a climax with the contemporary trend of public-private (or private-private) granting or Internet auctioning of infrastructural equipment or even complete infrastructural installations. This approach turns the notion of value extraction from the assets (dear to Asset Management thinking) into a mere calculation on instantaneous cost (contrary to the notion of life cycle) on a deal settled for the lowest bidder. We should even point out that some agents in the business seem to be very pleased with this trend, despite all its inherent flaws.

It is undeniable that firm control over procurement costs constitutes a crucial advantage for maintaining the financial health of an organization endowed with a heavy capital-intensive assets fleet, both in the industrial and infrastructural sectors. Indeed, purchases represent the equivalent of up to 80% of the organizational turnover in sectors such as the automobile or the distribution industries. It is no wonder, thus, that the price of the assets in which they invest appears as such an essential factor for industrial leaders worldwide. However, all too often this prevalence of the economic aspect tends to overshadow other trends that should also be considered—most notably, that of reliability. We are faced here with a certain organizational culture, very widespread nowadays, which has shunned all strategic and systemic vision in favor of a restrictive short-termism.

As we've already suggested, an exemplary purchase should respond to a definite need at the lowest possible overall cost, rather than at the lowest possible price. It should be noted that according to this definition, the procurement price is a secondary factor. However, it is the most prevalent in real-life procurement decisions, whatever else one would be inclined to believe. This rationale has led countries in which production costs were the lowest to gain situations as the "factories of the world," for the simple reason that these states spoke the language of industrial leaders: that of the lesser cost. This geographical displacement of global production has generated a remarkable and generalized qualitative degradation of assets. Indeed, one should consider that the production realized from the assets depends on three modulable parameters: namely cost, timeframe, and quality. However, it is an often neglected fact that it is only possible to modulate simultaneously two of these distinct parameters. When buyers prioritize assets produced at the lesser price, it is only logical that the machines' quality should decrease accordingly. Caught in a twisted game of which "minimal investment" is the golden rule, managers—the internal clients of the buyers—are bound to go from disappointment to disappointment as they are increasingly resigned to do without efficient machines (in the Asset Management understanding of the notion).

By solely focusing on prices, the buyer favors his own interests; he strives towards the optimum of the "procurement" function to the detriment of the needs of other segments of the organization (and all too often, he does so in good faith). Yet it is obvious that these segments, which themselves heavily depend on the buyers' decisions to reach their own realization, must be taken into account at the time of the procurement-related decision-making. Too little consideration is given to the well-informed opinions of engineers and operators—in fact, buyers and organizations often reenact purchases previously denounced by these agents. Hence the input of life cycle costing, and of a quickly-developing array of methodological tools. By monetizing the financial impact of an asset across its entire life cycle, one obtains an increasingly precise notion of the purchase's profitability, as long as one considers its entire existence within the organization. However, design-to-cost remains the predominant method for assessing costs, whereas life cycle costing is still a marginal trend, or at the very least applied in a timid fashion—this will be developed in coming chapters.

Agents of the "procurement" function cannot take the entire blame for this situation. From the perspective of enterprise sociology, the so-called "buyer's mindset" is in fact determined by exterior factors such as corporate pressure or the lasting influence of a culture that rewards quick successes. Nonetheless, if this explains why the buyer cannot be an internal consultant (since he is not required to heed the needs of other segments), he should at least learn from his mistakes—which rarely seems to be the case. De facto, mediocre assets are becoming the norm within the industrial sphere, and operational expenses (OPEX) are on a rapid rise. The OPEX we are discussing are not those that are originally budgeted, but those that are eventually spent throughout the budgetary year, not to mention the shortfalls tied with unplanned production stoppages; these are such tremendous wastes of funds!

Buyers are buying irrationally—and quite stubbornly so. One cannot help but to think of Albert Einstein, who defined insanity as the act of "doing the same thing over and over again and expecting different results."

When one engages a reflection of procurement processes, it is essential to discuss the question of the regulations imposed on the public markets. Indeed, we could synthetically assert that the public market regulations in most countries dictate that procurement be realized at the lesser price, following the cheapest bid. Disregarding the reasons supporting this present condition of the public market, it is quite clear that if the economic world embraces Asset Management in the near future, these regulations will quickly become totally obsolete.

To support this assertion, one could enumerate a number of points:

- What is known as the "negative elasticity" between low CAPEX and the consequently very high OPEX in the remainder of the asset's life cycle. In other words, when equipment is bought for a period of at least 10 years in this fashion, the probability of not enjoying a functional and qualitative operation of the asset rises spectacularly. Allow me to detail this claim by providing an example closely tied with the concept of LCC: in the public transportation sector, it is common knowledge that by the time a piece of rolling stock is decommissioned, the value of the initial CAPEX will have been spent somewhere between five and seven times in OPEX (in updated currency). This ratio is quite similar in any other industrial or infrastructural sector. It is therefore clear that procurement at the cheapest price induces a harmful evolution of OPEX throughout the life cycle. However, intellectual honesty demands that we also point out that the injection of a CAPEX supplement in the procurement phase does not suffice to ensure the correct management of OPEX over the rest of the life cycle. As we know, it is the best-practice conduct of the procurement project (considering both what is integrated in capital expenditures and the anticipated expectations of OPEX performance) that will indeed generate desirable results.
- The contemporary regulations inducing "cheapest bid" deals for public markets are inherently unable to avoid the pitfalls for which they've been designed; if they could, we would know it by now. Indeed, their enforcement is not (or not anymore) an obstacle to the occurrence of conflicts of interest, or even of events of corruption; once again, we would know it by now.

One can only rejoice in the fact that some countries have really seized the opportunity to tackle this problem, and have succeeded in developing functioning models of public markets aligned with the fundamental principles of Asset Management, but not solely—they are especially very supportive of states that act as financial investment backers using public funds. New Zealand, in particular, has for the last decade led a successful campaign aiming at the redefinition of these public market regulations, taking a clear stand in favor of the consideration of the life cycle as early as in the procurement phase. Furthermore, it has operated this transition in great simplicity: the public market regulations in New Zealand state that the replacement of assets reaching the end of their life cycle must be priced by the assets which will replace them at the time of dismantlement—and that this must occur as early as in the purchase offer stage. This implies that the offer should integrate a global life cycle cost attractive enough for the renewal to be profitable for the buyer; hence, the system encourages a long-term sight by making bidders project a vision on the end of the life cycle and help their clients in the efforts they will have to deploy to renew the asset.

In practice, this implies that the stated requirements should include a tender amount that will integrate the initial CAPEX as well as the CAPEX provisioned for the identical renewal during the end of life.

Other countries are nowadays trying to draw inspiration from this model, not only in Australia but in various other parts of the industrialized world. This trend is all the more laudable if we consider that, historically, other attempts to include the notion of "life cycle" in public tenders had systematically failed. As an example, we could state the case observed in the United States in the 1980s. The Department of Transportation at that time, encouraged by the wave of liberalization and privatization of public infrastructure implemented by the Reagan government, had developed public market specifications whose only admissible tenders in the urban transportation field were those in which the life cycle cost of the equipment was eligible; in other words, instead of favoring the lowest procurement CAPEX, this initiative offered to award the markets to the lowest LCC. These clauses were successfully imposed for a few years in the United States, but they were abandoned as soon as the state agencies became aware that the bidders were in a situation of such ignorance regarding their LCC projections that the purpose of the initiative was fundamentally discredited.

To conclude on a positive note, we could assert that despite market regulations remaining well under the standard set by Asset Management principles in most countries, undeniable progress is being made in the present era. The emergence of a dosage between price, technical quality, performance, and HSE criteria will potentially bring into question the habit of buying at the cheapest price, even if this may not suffice to induce a true taking into account of CAPEX and OPEX throughout the life cycle.

2.2 Outsourcing and Accountability

Over the course of the last decades, a new trend has become the standard in the context of industrial contracts. In addition to material purchases, organizations now purchase an increasing quantity of services. Hence, entire sectors of industrial and infrastructural management are currently being delegated to outsourced companies—the most visible part of these transfers has probably applied to IT services, but they are also happening in the industrial maintenance sector.

The market share of outsourced industrial maintenance is in constant augmentation. This means that organizations have become aware that they could delegate those specific tasks that external agents could realize more efficiently and at a lesser cost. This practice has encountered such success that industries have gone a step further, transferring ever more of their activities to outsourced companies in order to attempt to simultaneously transfer the responsibilities inherently tied with the risks. However, the standards that apply at the moment state that only the corporate director may be deemed accountable for risk; hence, we can see through the illusion embodied by the "performance contracts" whose results are made up of parameters that may by no means be tied with a subcontractor's responsibility, in terms of social or legal accountability. This approach is therefore practically limited in terms of performance, due to the fact that organizations lack a normative frame suited to their wishes.

The ISO set of standards (in our case the ISO 5500X) have emerged in this complex context. They have acted as a true game changer by integrating within their management systems the notion of *outsourcing*. The ISO standards' approach consists of defining a frame in which a free but aligned practice of outsourcing may be exercised. In short, one can consider that organizations who seek to obtain the ISO certification are entitled to purchase and outsource any service or product, so long as they ensure that their management systems are totally aligned with their subcontractor's (and vice-versa). Thus, these standards regard outsourcing as an integrated extension of the organizations' management system within which the alignment of operating modes is the foremost priority. It is therefore required that the subcontractor take part in the virtuous circle realized by the organization in its process of continuous improvement. These ISO standards mark a first external effort to implant more systemic and sustainable conceptions in the procurement sector.

It is also necessary to discuss the inherent subjectivity, which is tied to the procurement function and which comes with a heavy consequence: that of divergent appraisals. I recently overheard an old friend, himself a purchasing manager and a specialist of the supply chain, assert that the industrial world is faced "with a real problem when it comes to assessing purchases because there are no buyers' schools," or at least, not in the sense that there are "engineer schools" or "manager schools." There are, in fact, a few specialized degrees and MBAs—but we must admit that this role is all too rarely regarded as a fully legitimate segment on its own in the sociological representation of organizations. Furthermore, the buyer is characterized by his role as an "interface" between the operator (who employs him) and the suppliers with whom he is asked to negotiate. However, it must be understood that these two tribes do not necessarily speak the same language. The operator knows that he needs a performance but he does not necessarily master the expertise required to evaluate the quality of the product or the service that was provided; on the other hand, the manufacturer may not exactly understand his customer's requirements, or worse yet, he may understand them but be unable to come up with a satisfactory solution.

Hence, it is not necessarily easy to give long-term guarantees to the purchases made, which is a real problem. All of these notions reinforce the idea that a departitioning of the procurement sphere is in line with the empirical reality.

My professional activity allows me to visit over 20 industrial sites every year. In some of these sites, I've witnessed truly ludicrous procurement systems that have caught my lasting attention. Indeed, I've seen a number of cases where the buyer's salary was indexed on the difference between the maximum and minimal offers that they were able to deal. Imagine a procurement function whose remuneration depends on fluctuating variables, and tasked with looking for the lowest value within the broad set of offers that it has at its disposal! It is a truly suicidal policy: not only does it entirely neglect the assets' life cycles, but such circumstances make merely bringing the machines up to their commissioning unlikely! In other cases, phenomena of corruption add an extra layer to the perversity of this system: one can then observe, for example, a recurring usage of preferential purchasing despite the absence of any strategic rationale to account for these decision-making mechanisms. Hence, the illusion of a healthy competition between qualified contestants during tender processes is clear, since this function generates results completely opposite to those of its primary purpose.

In quite a different register, one could also discuss the cases of concession infrastructures, or of PSD (public service delegations). In those concessions or PSDs, the operator and maintainer is not inherently the assets' owner. The owner delegates the procurement function to a third party with whom he establishes a contractual agreement. This practice generates additional issues in terms of procurement rationales and costs ponderation. The first transfer (from owner to buyer) entitles the owner to include clauses regulating the taking into account of purchases over all of the concession duration, and even beyond their "period of accountability," as it is defined in paragraph 4.3d of the ISO 55002 standard.

Indeed, both in the context of an industrial organization that owns its assets fleet and in that of a concession, the ISO 55001 standard applies to what it designates as "the accountability period." Thus, there is a hint of bad faith in asserting that outsourcing, delegation, or concession are not so many fields of application of Asset Management principles. Moreover, a concession is eligible for ISO certification—in fact, concession certification occurs on a daily basis. There is therefore no requirement of ownership to act in accordance with the frame defined by the Asset Management standards.

2.3 Changing the Terms of the Procurement Function

We have already asserted that the alignment of objectives and needs is a central factor in the improvement and the reinvention of the procurement function. Indeed, it is a requisite for the optimization of the usefulness of this segment, in the sense that it would strengthen its roles of supervision and innovation. The act of purchasing must become a major factor in the extraction of the assets' value, rather than to remain the product of a punctual negotiation in time and space. One can easily envision how Asset Management could facilitate the path towards the self-improvement of this sector: indeed, the systemic and holistic vision which we promote could allow buyers to assess more thoroughly the impact of their decisions in order to learn from their mistakes and to cap losses while optimizing the choice of suppliers and machines so as to fulfill the expectations of each of these segments—at a minimal LCC rather than at a maximal CAPEX!

Of course, one cannot simply assert such global propositions without offering a more practical perspective. It is interesting to consider the techniques and capabilities inherent to our discipline and whose input in terms of procurement-related decision making could turn out to be decisive.

Thus, one cannot overstate the importance of implanting the CAPEX/OPEX trade-offs within the buyers' mindsets and work methodologies—or in other words, of exposing the disproportionate weight of operational costs relative to the initial price in the context of an asset's life cycle. One should keep in mind that the average asset will have cost, by the time it reaches its end of life, an OPEX representing three to seven times its original and updated purchase price (CAPEX).

One could legitimately wonder why buyers consistently act as though they were unaware of this statistic. In fact, the answer is as simple as it is perverse: they are not necessarily aware of the reality of the assets' life cycles beyond the procurement phase. Furthermore, CAPEX are funded with shareholders' capital, while OPEX are funded with the organization's production revenues. This system therefore encourages procurement managers to buy at the lesser cost (in order to satisfy shareholders) while turning a blind eye to the widely superior expenses which will derive, in the middle and long terms, from this decision.

This is the lesson we learn from risk/cost and CAPEX/OPEX trade-off calculations; the more one seeks to buy at the lesser cost, the more one exposes himself to the increased risks which sooner or later will induce recurrent costs infinitely superior to the initially realized one-off savings.

Trade-off methods are nowadays universally available. It is urgent that industrial actors cease to consider that these calculations depend on a particular and confidential expertise detached from day-to-day realities; on the contrary, they allow for concrete and sustainable savings. One can wonder at the idea that purchasing departments have not yet taken more inspiration from these systems, rather than to go on perpetuating a counterproductive approach which shows little success in creating value for the organizations. One cannot overstate this particular point: it is essential that purchasing professionals and organizational management realize that the tools and techniques required to implement a procurement system enabling an optimal extraction of value from the assets are at the disposal of every organization in the modern world.

Hence, one can easily grasp that the essential value that buyers should be taught is that of alignment (a notion inherent to Asset Management and the ISO 55001). It is this alignment that will make possible the necessary evolution of purchasing departments. Indeed, in the context of an aligned procurement policy, one is able to cap losses and useless purchases by the means of an increased interdisciplinary communication (via the Asset Management System). Furthermore, costs will drop as the number of suppliers decreases—indeed, one can observe that for products of equal value, costs are proportionately higher where there are more numerous suppliers. In addition, switching from one supplier to another comes with risks that purchasing departments have difficulty monetizing but that can turn out to be very heavy.

Furthermore, reducing the number of suppliers means improving the connection with those that one continues to do business with, and as we've already suggested (without falling into the pitfall of rigid dogmatism), all that tends towards sustainable exchange falls in compliance with the principles of Asset Management. One can observe that the progressive democratization of the recourse to KPIs (which allow for the implementation of continuous improvement measures implicating both customers and suppliers) is inclined to favor the kind of alignment promoted by Asset Managers.

Asset Management professes that organizations should align their procurement systems with what the ISO 55001 refers to as the SAMP (Strategic Asset Management Plan). This element constitutes a decisive factor of risk ponderation in the event of a change of supplier. Therefore, it is essential that one be able to correctly *monetize* risk and to represent its *trade-offs*.

To conclude, it is useful to approach the problem of procurement systems from a more sociological angle. One can observe that in stark contrast to the isolated activity of procurement as it is practiced in the overwhelming majority of the industrial spheres, the aligned vision promoted by Asset Management implies an important evolution of the buyers' mindset, but also a reevaluation of the managers' expectations, since their judgement will always influence the buyers' decisions.

I believe—and wish to believe—that Asset Management offers a new approach of procurement, by opening up the latter to new factors and problematics. In the long run, our aim is to demonstrate that purchases, just as well as any other segment of the assets' life cycles, are a part of a more global function. Obviously, purchasing is not maintaining; however, the purchased machine will have to be conserved. It is both absurd and ruinous to dissociate the purchasing from the operational. The interdependency of functions and performances is immense, and it can no longer afford to be ignored by the industrial culture.

Let us state once more that what is truly sustainable (responding both to functional needs and anticipated performances) may by no means be determined by a "cheapest bid" procurement criteria. We have much to learn in this regard from the positions promoted by environmental and sustainable development experts who choose to analyze the ecological issue through the prism of the obvious contrast between decisions aiming for short-term (apparent) benefits and their destructive planetary impact on the mid and long terms.

This issue brings us back to the question of "value," a prevalent notion in the field of Asset Management but one that is rarely developed in a systematic fashion within organizations, even though their survival depends on lasting strategies rather than on annual dividends.

CHAPTER 3

FEED, RAMS, and Cost-Effective Availability

Maintenance professionals (and, by extension, the actors of Asset Management) have always been faced with two sources of concern, which paradoxically appear both contradictory and complementary.

The first of these concerns is tied with a form of susceptibility inherent to our activity: since maintenance practices as they exist today throughout the industrial world remain costly, hazardous, and often unreliable, the maintenance operator blames a system which does not recognize the value of his expertise and contributions. Since he is aware of the processes that could diminish the assets-related risks, the operator resents management for not requesting his input in the earliest stages of the machines' life cycles, and therefore not allowing him to stamp out the reproduction of a pattern whose inefficient and anti-economic character is well-known to him. Hence, the operator feels a form of vexation and jealousy toward the designers and buyers who conceive their strategies from the comfort of what he sees as their ivory tower, based on fatally incomplete information.

We should keep in mind that the maintenance sphere's attitude is quite natural; it originates in the frustration resulting from being systemically called upon for contribution too belatedly in the process, and from still being regarded as a "parts-replacing function," which can generate frustration among qualified engineers.

This first concern is typical of a deeply-rooted situation, which is fortunately beginning to lose ground. Progressively, and in a rather heterogeneous fashion, the industrial world has finally begun to realize the strategic input embodied by the maintenance staff's expertise as early as the preoperational phase. In the last 30 years, industrials have therefore begun to request the opinions of maintenance operators. This, in turn, has been the cause of our second source of concern: *will we truly know how to help them?*

3.1 The Input of Front-End Engineering and Design

During the noughties, industrial actors took notice of a new trend: FEED, or *front-end engineering and design.* This international protocol was innovative in that it allowed engineers to take into account voluminous and diverse information upon the development of their design processes. In fact, this approach has permitted the implementation of a new relation to the produced assets, based on principles of alignment: for the first time in such an effective and formal manner, the process facilitates foresight and projections on the operational life.

Naturally, FEED brought Asset Managers so much unprecedented data that they unconditionally adhered to its practice. However, even if this new approach headed in the direction of our way of thinking, we knew that this global vision on different factors such as design, risk analysis, or CAPEX balance was still rather blurry despite being more informed than all previous approaches. Indeed, the FEED methodology was based on a process that was all too sequential, too strictly mechanical, and it was therefore only seemingly systemic. In other words, the methodology had a holistic and exhaustive ambition, but it did not derive from an organic process. However, this new and successful approach proved useful in the sense that it helped make industrials (or at the least, the more innovative industrials) realize how predominant a factor foresight was. Better still, some concrete applications succeeded in generating remarkable total savings and in opening up real perspectives of new value extraction.

From this observation emerged a number of problems, which I feel it is important for us to reflect on:

- How should we profit from this significant breakthrough, while demystifying the elements of operational maintenance that were not taken into account until that point?

- How could we mitigate the *susceptibility of the maintenance operator,* since we were now working along the lines of the considered and preventive approach that he had promoted until that point?
- Finally, how could we go beyond the FEED method to reach a truly holistic system?

The success of the FEED approach among operators was therefore set to create an unprecedented situation, which would now be to our advantage. The public opinion was finally realized that an informed anticipation of the asset's life cycle was truly necessary, and we were determined to make the best of our newly found importance. Thus, it struck us as crucial to rethink and improve the FEED processes in order to make them comply with our systemic ideal.

3.2 The Limitations of the RAMS Approach

All too often, the inclusion of information linked to the operational phase has been based on RAMS (Reliability, Availability, Maintainability, and Safety) studies. RAMS consists of a series of risk and vulnerability studies that take the form of an exhaustive set of anticipations focused on these four dimensions and launched as early as the design phase. This process has come to be known as Risk Management, and its development has occasioned a number of international sectorial standards, which in turn gave way to the definitive standard regulating Risk Management systems: the ISO 31000.

It would be ill advised to blame industrial leaders for such a practice, which values a preventive, and therefore foresighted, approach to operation, logistics, and maintenance. But too often, one can observe that RAMS has become a systematic and quasi-official response to the formal requirements of industrial customers, who allegedly refuse to begin construction before the supplier can provide them with a RAMS study. On this matter as well, we could celebrate their judgment. But that would mean turning a blind eye to the fact that these documents have become the industrial equivalent of medical certificates—stately, quasi-magical documents providing operational warranties as early as the construction phase. It has therefore been observed that, having become practically compulsory, these studies were produced "by the mile," without being the object of any particular reflection beyond the formal imperative to complete the document. Over time, these studies have become constraints for engineering departments, even if the professional "Omerta" forbids agents to expose their lack of input; prohibitive and less than engaging outside the sphere of contractual bureaucracy, they are moreover undertaken too belatedly in the operational schedule.

Furthermore, these studies are eminently technical, as opposed to economic. The reflection on economic optimization is therefore doomed to only be undertaken separately, and retrospectively. This is an indicator of a serious flaw in the systemic alignment.

At the time when I was managing an urban rolling stock network, I was totally unaware that RAMS studies existed for every one of the assets we operated. I can assert from experience that my case was by no means isolated. Quite often, operators are still oblivious to the fact that such documents even exist! Worse yet, even those who know of their existence are unable to use them to any profit; indeed, these heavily technical documents, intended for engineers savvy in the field of reliability, are particularly hermetical. Once again, one can denounce a misalignment in the language practiced by the different agents of the asset's life cycle.

Yet, these approaches have an undeniable value in terms of rigor and methodology, and they represent a remarkable input in terms of collected information. They are, indeed, necessary—but they are also poorly harnessed, since they are not realized in the "language" of operators. As things are, they serve solely for engineers, and as such, they are less useful than they should be, largely because of a nonalignment of capacities.

While they remain serious studies, they can therefore provide only a relative input for operators dealing with arbitrations in the project phase. Hence, these approaches considerably restrain the overall Asset Management performance:

- A certain number of risks are not taken into account (or decision makers tolerate inconsiderate risks due to non-alignment).
- Furthermore, because these studies are not inscribed in a process of economic optimization, decision makers end up agreeing to prohibitive prices or making inexcusable compromises over the reduction of risks.

The situation is paradoxical to say the least: despite the fact that FEED techniques are now broadly regarded as crucial from a formal perspective, their input is still vastly disregarded by most operational agents. Operators continue to hire, on the eve of commissioning, employees who are content to follow word for word the suppliers' instruction manuals. Hence, important and available information is consistently not put to profit. Therefore, the functional analyses that constitute the primary value of FEED are totally neglected at the time of entering the operational phase. This is an inexcusable waste of resources that could be transformed into so many opportunities for success.

The following factor is particularly important: we see that upstream, viable methods that take into account projections on operation are being implemented, but since the operators themselves are not aware of these efforts, they persist in reproducing the usual ultra-mechanical and reactive maintenance schemes.

What emerges from this situation is that the actual state of things satisfies neither of the mobilized parties. Design managers, who effectively realize the studies, do so under a form of constraint; they know that RAMS do not guarantee an optimal performance in operation since the roots of the studies lie in the preoperational phase. These studies are useful for the hand-over but those who operate the assets are not able to profit from them; in short, RAMS makes it easier for one to sell a machine, but not to use it properly.

On the operator's part, these studies are almost impossible to use since they do not suffice to obtain guarantees, which are necessary as soon as one seeks to establish informed and safe maintenance and life cycle management strategies.

It is in this somewhat chaotic context that the crucial problem of cost-effective availability intervenes.

3.3 Cost-Effective Availability and Optimization: The FEEDMAN Protocol

In my time as an operator, I was quick to realize that maintenance managers were only preoccupied with the "availability" factor, to the detriment of a reflection on cost-effectiveness in a longer term. Hence, I focused my efforts on availability, in order to reach the standards set by management, and I was congratulated for effectively reaching the desired objectives in terms of availability. It was only in hindsight that I realized to what extent I'd been performing an incomplete labor by aspiring solely to guarantee a level of availability at any cost. Far from being trivial, this corporate obsession with the availability factor is in fact a considerable fault, whose impact on the current practices in the operational phase cannot be overstated.

Indeed, in order to effectively generate availability, one has to be able to ensure that the undertaken effort generates an availability that costs less than it yields. This seems to stand to reason. However, operators often turn a blind eye to the expenses tied to the mainly corrective maintenance they practice, as these expenses always seem less heavy than those that would come with a preventive-orientated organization of their plant's maintenance. The decision makers' choices to strictly focus on reaching a maximum technical availability makes them oblivious to the fact that this objective may not be satisfied without high levels of expenses. One should also keep in mind that every operational system has its share of cost-effective availability and its share of senseless availability. Insidiously, these two antagonistic (in terms of competence) types of availability soon morph into multi-color graphics in the "Dashboards" and become more dazzling still when they reveal a new "Record."

This trend, which is deeply rooted within our corporate mindsets and places availability at the center of the issue, is therefore inherited from a distorted vision, and nothing guarantees that it may result in the elaboration of a successful strategy. Productivity, relative to the profitable share of availability and distinguished from the unyielding share of productivity, is the factor through which profit is realized. From the perspective of value extraction, economic value should only be conceded to the profitable share of availability. This is one of the issues on which the worlds of maintenance and Asset Management can agree. RAMS studies neglect this distinction between the different types of availability—hence their shortcomings.

However, if the operator manages to realize the RAMS studies in a holistic and aligned perspective, guided by the principles of Asset Management, then he will benefit from an unchallenged insight over the profitability of availability.

It should be kept in mind that FEED was designed by major engineering offices who needed to bring some extra structure to their design and engineering approaches. On the other hand, RAMS was crafted by designers, following a radically different purpose: instead of facilitating the design phase, it aimed at a clearer assessment of technical constraints.

The idea behind the FEEDMAN process was to operate a critical analysis of RAMS studies through the prism of FEED inputs. Indeed, we were quick to realize that FEED was based on RAMS notions (notably the famous Reliability Block Diagrams, which create a *functional structure*, which in turn allows for the definition of a functional availability). The challenge that FEEDMAN set out to accomplish was to merge the FEED approach (which takes into account notions of reliability) with an original idea: not only to try to operate a simulation in terms of availability but also to integrate CAPEX/OPEX trade-offs, and therefore a perspective of *economic* certification. It is, therefore, a means of promoting a simulation of the risk/cost factors as early as in the preoperational phase.

Indeed, boasting an approximation of the profitable availability is the soundest method of avoiding the frequent pitfall, which consists of assuming that "economic optimization will be taken care of in the end," or simply "later," as we so often hear.

Historically, the specifications received by the design offices were given in terms of availability and reliability. In this case, however, designers are asked to state the budgetary constraints much earlier in the process. By including notions of foresighted maintenance and operational economics in the earliest stages of the preoperational phase, one is able to shine a light on the *profitable* availability. The issue here is indeed to monetize the risks tied with each factor implicated in the machine, in strict alignment with Asset Management policies. This approach allows designers to guarantee more reliably the performances in the operational phase and in the end of life.

3.4 Key Factors of the Profitable Availability in the Project Phase

We can outline three major key factors allowing one to operate an anticipatory and efficient verification of the profitable availability of future installations:

- Envisioning the operation and maintenance of installations by sizing the operational and managerial effort through the prism of the consequences over the functions that the equipment should fulfill (this is known as the system approach) rather than solely through the failures that might occur (equipment approach). The use of the RCM methodology (Reliability-Centered Maintenance), if it is applicable with a vast economy of scale, is part of a rationale seeking to map the installation's productive processes and functional groups in order to list the more pertinent maintenance tasks relative to the given stakes and to prioritize the economic effort of maintenance based on tangible and demonstrable techno-economic vulnerabilities. This approach to constructing Asset Management decisions marks a decisive qualitative leap in the field of maintenance strategies, while preserving the intrinsic reliability of equipment in a coherent and rational manner relative to economic imperatives.

- Sizing strategic stocks for the sustainability of production, relative to their risk/cost relations. This approach is nowadays regarded as the only one allowing the definition of the pertinent volume of spare parts while respecting the balance between the risk of parts' unavailability on the one hand and the cost of fixed capital in acquisition on the other. Originally applied to existing installations (due to its obvious pragmatism), this method is increasingly and successfully deployed on the first endowment of a project's spare parts. Apart from its input linked with the sizing of the pertinent volume, its applications in the procurement phase of major industrial programs provide a new guarantee against the historical controversy regarding parts supplied in packages by the manufacturers with the equipment itself, and which gives operators (or maintenance agents) the unpleasant impression of paying for unnecessary parts. Indeed, thanks to the risk/cost simulations it has become quite possible for industrial buyers to determine simultaneously the amount (in %) dedicated to first endowments stocks and the unacceptable vulnerabilities that allow for clients to realize an objective definition of the parts packages supplied by the manufacturers.
- Operating a definition of the preventive maintenance periodicities and/or production stoppages that focuses more on quantitative metrics than on purely qualitative figures (as is most frequently the case). By cali-

brating the risks and the costs relative to time or to the production unit, the economies of scale linked to these intervention periodicities will become more pertinent and more adapted to the operational reality than the logistical criteria of design offices. Furthermore, this approach provides an immediate connection with RCM studies and facilitates the optimization of strategic stocks.

It is important to underline the fact that the practices associated with these key factors are inherently aligned with the concepts of Asset Management and are therefore so many catalysts for the obtaining of the profitable availability targets.

From a designer's perspective, implementing a redundancy would suffice to lower the risk exposure, but that would mean neglecting the cost, which would be doubled! The FEEDMAN protocol allows us to highlight such issues.

These three key factors are therefore the founding pillars for FEEDMAN, which we will describe in the following subsection.

3.5 FEEDMAN—FEED Aligned with Asset Management

The FEEDMAN idea integrates (hence its name) two major concepts: "Front-End Engineering and Design" (FEED) and Maintenance and Asset Management with the techno-economic performances to achieve on the short, mid, and long terms.

The starting point of the FEEDMAN approach is the same as that of the RAMS studies; namely, the will to operate a provisional verification of availability. Based on functional models and on the installations' production processes (Figure 3.1), the modeling of the Reliability Block Diagrams is then produced. After the allocation of lambda and mu performances (and not solely of the MTBF and MTTR, as is often observed), the simulations of productive availability occur (Figure 3.2). The targeted objective is obviously to identify the unity, subsets, and equipment, which present a poor factor of contribution to the obtaining of the targeted productive availability (the infamous *bad actors*).

Up to this point, the approach remains quite similar to the well-documented RAMS studies, even though we can observe a number of differences in terms of the processes used to acquire the operational information, particularly in the use of elicitation.

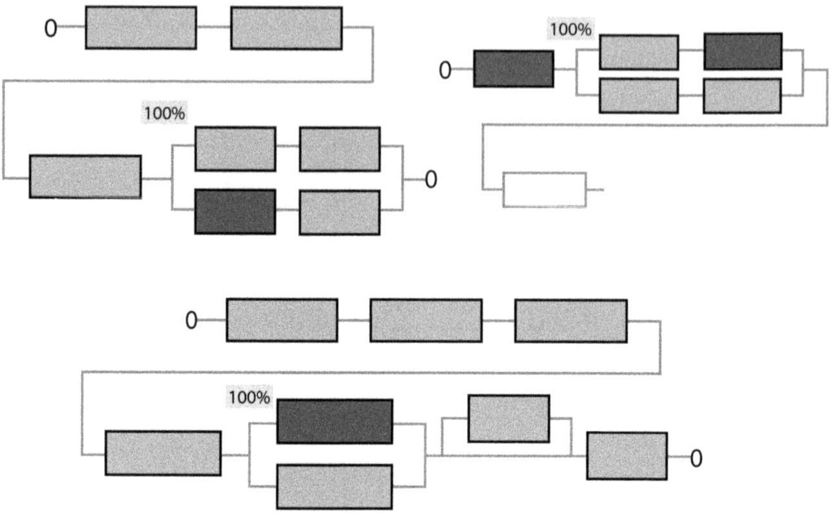

FIGURE 3.1 Model for provisional availability verification

UNIT	ITEM	% Loss/Year	Failure Frequency/ Year
SMR	Combustion air fan 07B01	0.33	0.66
SMR	Flue gas fan 07B02	0.33	0.66
HDS	Reactor heater 11F01	0.20	0.50
HDS	1st HDS reactor 11D04	0.15	0.07
HDS	2nd HDS reactor 11D05	0.15	0.07
SMR	HT shift reactor 07D02	0.15	0.07
HDS	Recycle compressor (+ motor) 11K01R	0.12	1.26
HDS	Recycle compressor (+ motor) 11K02R	0.12	1.26
SMR	Desulphuriser 07D01	0.11	0.34
HDS	HP amine absorber 11C01	0.11	0.34
HDS	LP amine absorber 11C04	0.11	0.34
HDS	Vacuum dryer 11C03	0.11	0.34
HDS	HP separator 11D06	0.11	0.34
HDS	LP separator 11D10	0.11	0.34
HDS	Dryer overhead separator 11D12	0.11	0.34

FIGURE 3.2 The major contributors to failures:
Representation of the simulation results that highlights the bad actors

It is only in the next stage that FEEDMAN's profound originality will appear for all its value, by providing a more sophisticated verification process that focuses directly on profitable availability.

By clearly distinguishing between good and bad contributors to the profitable availability, we push towards the realization of RCM studies those units who pose no threat to the accomplishment of targeted performances.

Beyond the proactive character of the task, which consists of defining maintenance plans in this early stage, the many feedbacks from which we profit clearly demonstrate that this practice is a strong asset when it comes to avoiding the pitfalls tied to an approach that seeks to define the maintenance logistics *prior to* determining the relevance criterion of its tasks. Furthermore, this practice also protects the future maintenance operator of the installation from the temptation of sizing his activity according to a "piece by piece" approach (Figure 3.3).

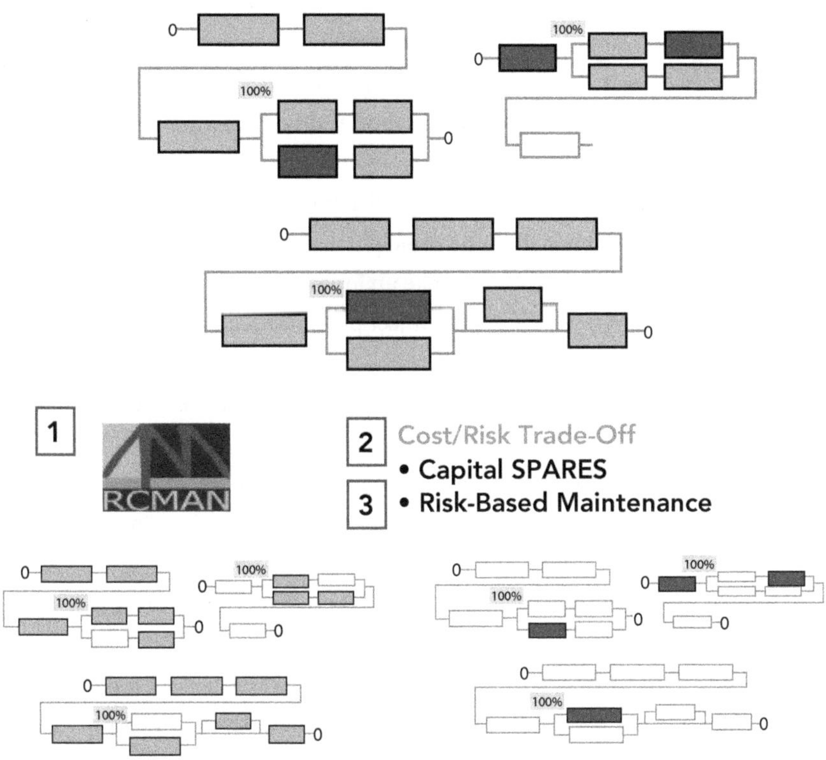

FIGURE 3.3 FEEDMAN's findings: Units subjected to RCM studies as soon as the FEED-phase simulations have been completed

Regarding the units identified as "bad actors," FEEDMAN provides risk/cost simulations allowing one to operationally make up for the shortcomings of the projected equipment in terms of reliability or maintainability.

The logic here is not to resort to traditional RAMS solutions, which would only seek to bypass the problem set by the performance shortages of the "bad actors" by offering to replace these actors with different technologies, equipment, and materials.

Very widespread in past decades, these replacement solutions are now rather obsolete since design offices develop installations of a very high technological level from the earliest project versions, which has a strong impact on reducing the leeway in terms of replacement or upgrading. Furthermore, the FEEDMAN risk/cost simulations are more efficient than traditional methods when it comes to discussing the traditional recourse to redundancies in the case of the identification of "bad actors."

Indeed, who among our ranks can claim that they've never witnessed the use of a redundancy for a "bad actor" equipment inducing another problem of performance shortage for what had then become a "double bad actor?"

In order to counter and avoid this unwelcome situation, FEEDMAN offers to review the risk/cost relations of the "bad actors" or critical units and equipment. This approach is based on the principle that it is operationally possible—under very realistic conditions—to make up for the high criticality of these units by implementing dispositions in the project phase, which directly implicate the organization of the operational and maintenance phase of these assets. Hence, FEEDMAN brings into play CAPEX/OPEX trade-offs and scenarios, which has the double advantage of, on the one hand, presenting the solution(s) in the face of the "bad actors" as so many project constraints, and on the other hand, placing them in the context of the techno-economic stakes of the entire life cycle of installations.

One of the possibilities that then transpires is that of integrating the risk of equipment unavailability in the strategic stocks simulations. This is one among various specific processes aimed at setting the level of risk tied with the absence of spare parts, to be established between the optimal level, which sometimes requires an unacceptable capital immobilization, and the current level, which usually does not take into account the economic impact of risks.

The approach elected to define the amount of parts for each item according to the consented risk principle consists of determining a suboptimal level of risk, which is not only realistic (in financial and technical terms) but which also guarantees an acceptable coverage of risks and of the quality of service required by production.

In other words, this approach induces a double optimization:

- For items in surplus: FEEDMAN allows one to define the acceptable limit of stored spare parts reduction without generating a significant increase of risk (cost of the lacking parts). This limit is translated by the individual code of the item and the quantification of the reduction to be realized.
- For items in insufficient volume: FEEDMAN allows one to define the acceptable limit of investments bearing on new quantities of stocks, which may induce a reduction of risks (or in other words: whose economic impact would be superior to the costs tied with the lockup of spare parts inventories) from the base study of each unit. This limit is translated by the individual code of the item and the volume of parts to be added.

Another possible approach consists of making up for the "bad actor's" non-compliant risk exposure through adjustments in terms of preventive maintenance periodicity, in full awareness of the risks and costs at stake. Indeed, this expedient form of compensation for the risks of frequent failures of the "bad actors" may not be undertaken lightly, and must by no means derive from trials and errors in the handling of preventive maintenance periodicities. On the contrary, it must proceed from a rigorous approach based on costs and risks simulation tools applied to every scenario of periodicity variation, thus making the most efficient alternative apparent.

The good and bad optimization practices enabled by the variations of the preventive maintenance intervals will be discussed further in Chapter 6—What Is the Operational Life?

Thanks to this type of simulation, the improvement of performance is obtained via a double positive effect. On the one hand, since we are dealing with a new periodicity that we know to be optimal in regard to the task's risk/cost trade-offs, it is clear that it provides an increased level of availability relative to any superior periodicity; on the other hand, and through the same process, we know that it induces a more attractive cost than any inferior periodicity. In other words, this approach realizes a very potent example of what we know as *profitable availability* (Figure 3.4).

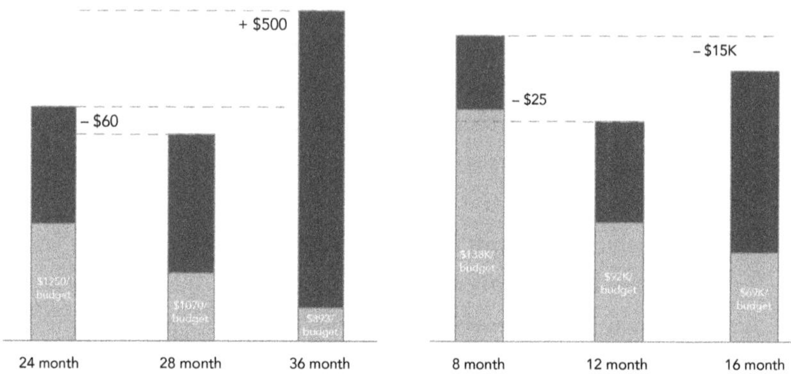

FIGURE 3.4 Arbitrations on maintenance intervention periodicities

By conforming to these orientations, the industrial designer implicated in the project phase anticipates the issues that would sooner or later have tormented the maintenance agent during the operational phase, with the considerable advantage of not "having to wait for the fire to occur to wonder how to extinguish it."

"Avoiding the fire," or, in other words, adopting a perspective bent on foresight and the anticipation of risks and costs, is part of the Asset Management process aimed at organizing a better value extraction from the assets.

From the perspective of a designer, cutting risks could be as simple as, for example, implementing a redundancy; however, that would mean turning a blind eye to the expenses, which would then be multiplied by two! FEEDMAN allows us to shine a light on these phenomena.

Integration of the operational phase in the project phase:

- Importance of capital spares: If one is unable to make the indicators turn green in terms of increasing the reparability.
- PM interval optimization: If, as an operator, I must take notice that things are as they are, I should first be concerned with the relationship

between risks and costs. In order to do so, it is necessary to optimize the relationship between risk of failures on the one hand and costs of maintenance on the other—or, in other words, to optimize periodicities.

Regardless of the quality of the applied methods and concepts, a successful transition from the preoperational phase to the next ultimately depends on the existence of a reliable organizational system. Only through the alignment of objectives and aspirations can a fruitful implementation of the lessons inherited from the project phase (and retroactively, from the operational phase) be achieved. One could indeed wonder how, in spite of the amount of technological tools available today, we can still observe such imperfect or failed transitions into the operational phase. That would mean neglecting the fact that the methods at work are based on a human and organizational motion that can be referred to as a "handover" between agents from the same organization.

Quite often, the corporate mistake is to envision this transition as a moment of rupture or as a turning point. Indeed, it is illusory to imagine that one can complete the project phase and begin to plan for the operational phase. On the contrary, the vision that must be implemented is that of a continuum: It must be a vision promoted by leadership. Success has therefore as much to do with the applied methods as with a corporate will to establish the fact that operation and design are two sides of the same coin.

When one reflects on the various paths that Asset Management has set for itself in its numerous embodiments in the industrial spheres, it becomes clear that the fundamental spirit of this field lies in a culture of the organizational. The crucial issue is always to implement a systemic and holistic perspective, focused on the whole of the assets' life cycles. Without this approach, it is most unlikely that the preoperational/operational interface may be efficiently overcome. In fact, paraphrasing Albert Einstein, it could be asserted that the problems of systems are not raised by these systems themselves, but that they're set by the interfaces between them.

In other words, it is difficult to approach the operational phase in full confidence if one has not yet grasped the weight of alignment in the elaboration of a successful strategy.

PART II

BUILDING ASSETS AND PREPARING THEM FOR THEIR LIFE CYCLES

CHAPTER 4

From the Preoperational to the Operational Phase

Construction and Installation

The issue at hand here is to insist on the risks and the stakes tied to this often-neglected segment of the assets' life cycles. Indeed, undertaking a serious effort during the design and project phases does not guarantee that the initial plans will be realized as such in later stages. However, of course, in order to accomplish the task that it has set out to accomplish, a given organization needs a set of tools and equipment of the utmost quality. Therefore, before operating industrial assets, one has to consider their construction, and to deal with the making process in absolute seriousness. Furthermore, it should be noted that during the physical installation of a plant or the implementation of an infrastructure or of a service, the organization must strive to optimize its dealings with various suppliers while making sure that the diverse components of what it is producing are both efficient and able to operate in a symbiotic fashion. Of course, one could observe that these issues are somewhat trivial for anyone who has ever had a foot in the industrial world. But once more, the devil is in the details.

4.1 Construction and Operations: Back to the Start

In order to give an appropriate illustration of the issue at hand, let us briefly consider the case of a renowned couturier who would recently have designed a dress that he believes to be his masterpiece. Between the project (his drafts) and reality, a considerable gap remains: The dress must yet be sewn and the design must be put to the test of realization. But indeed, a certain brand of optimism always pushes designers to assume that reality will mirror their ideas.

Studying the facts, one is quick to realize that the construction or the realization of a plan is an activity that must itself face a number of practical challenges and imperfections. Moreover, it seems fit to point to another limitation that is tied to the "silo-shaped" organization of companies. Indeed, since industrial organizations are excessively partitioned, the development of "psychological" attitudes specifically tied to the function of one of their segments is frequently observed.

Thus, one could speak to a project mentality that regards construction and installation as an end in themselves—such a mentality had already been observed when we were discussing the case of the procurement function. Of course, installation managers are not to take the blame for the rise of such a mentality: that which is to blame is an organizational system in which each segment deals with its own budget and objectives without striving for an increased interdepartmental cooperation.

In such a context, one can easily understand why the managers of each department would end up regarding the task they've been assigned as a goal in itself. However, we cannot overstate the fact that none of the segments that make up the asset's life cycle can be thought of as an "end in itself," since Asset Management only envisions the value of assets relative to their entire life cycles. We can therefore observe that almost every party involved agrees on denouncing the approach, which consists of leaving an industrial plant "ready-to-use." Indeed, what the operator wants does not come in the form of hurried physical readiness (e.g., a stainless paint job on the walls) but rather of a plant that would be ready to be used throughout its entire life cycle, or indeed that would already be inscribed in a rationale of long-term productivity and continuity.

One must only examine a plant's building site to realize that the metrics employed to assess readiness are "building-orientated," or, in other words, that the builders are striving to construct equipment that is functional immediately but not necessarily in the future. In the industrial vernacular, this phenomenon is known as the "handover" between the project phase and the Operational and Maintenance departments.

However, in the frame of Asset Management, this is a pitfall that must by all means be avoided. Of course, it is undeniable that isolated attempts at implementing more systemic approaches emerge here and there. But traditions are thick-skinned, and it is often difficult to make the organizational chart's different segments develop wider overlapping areas than those that they share today.

These issues derive from well-known pitfalls: the absence of a perspective bent on the entirety of the asset's life cycle (or even on the mid-term, which would already induce a welcome change of practice), the abusive pressure applied by installation operators, or even the egos of project managers who deny operators a say in preliminary arbitrations. However, one of the most prevalent issues identified when one reflects on the shortfalls of the installation phase is the lack of a sustainable operational mindset.

Yet, the operational corps doesn't fail to express the necessity of an evolution of this phase in order to make it more apt to take into account the specific needs of the other segments of the life cycle. And indeed, it would be a shame to pass on such a stellar opportunity to calibrate a number of mechanisms that will be much harder to implement later in the operational phase. In this regard, the installation can play a crucial role as a "lever" over the factors that must be harnessed. Hence, when one is setting into motion the operation of an industrial plant, one must tend to the recruitment and formation of a workforce, the preparation of the adequate structures, and the identification of the critical points of fallibility—all of which could have been more usefully addressed at the time of the asset's installation.

Moreover, operators frequently regret not having been able to more adequately anticipate these phenomena. Indeed, the teams tasked with proceeding to the installation are not expected to possess an operational expertise. One can therefore clearly perceive the usefulness of having a maintenance and operational workforce, which would already have been present in the earliest possible stages of the installation so as to ensure that the workforce is optimally aware of the assets' functioning. As it is suggested in the ISO 55000 chapter dedicated to *Asset Management Systems*, it is a shame to wait for a plant to be functional before thinking about supplying it with operators. All of this, of course, can seem to be pure common sense. However, because of budgetary constraints imposed on construction or of abusively tight schedules dictated by management and even more so by stakeholders, we observe that these practices are in fact very rarely implemented. On the other hand, operators must also be lucid on the subject of another pitfall linked with the formation of the employee's skillset: that which would consist of making the equipment *too* people-dependent, thus precociously encouraging the development of a "Loving Care" phenomenon (see Chapter 6).

4.2 An Opportune Moment

One should also observe that the installation phase is a privileged moment to proceed to the establishment of the Asset Register, or in other words, to proceed to the full and computerized arborescent census of the production area's individual assets, Asset Systems, and Asset Portfolios. Let us resort to a somewhat trivial analogy to exposit how this process can be not only useful, but truly necessary for operators from the very beginning of the operational stage. Let us therefore imagine that one aspires to build a castle made of hundreds of Lego parts. Rationally, it is easy to understand that it would be preferable to establish an inventory of these parts before proceeding to the castle's building, as long as it is still possible to distinguish each of the components that will be used in the process. Furthermore, if the installation phase is an ideal moment to establish the Asset Register, one should also keep in mind that it is profitable to use this time to realize the Asset Costing, since each part used will become an asset to be taken into account as CAPEX in the impaired asset's base. In addition to fulfilling the functions of an accounting register (which was already mandatory prior to the rise of Asset Management), Asset Costing provides operators with a correspondence between the assets' technical and accounting registers.

What a perfect occasion, therefore, to establish so early on in the operational phase the necessary and one-to-one correspondence between Register and Costing, since at this moment all of the parts in question are distinguishable and measurable!

This context is also exceptionally calibrated to engage in a thorough promotion of a common vision of the assets. If one succeeds in implementing a process in which financial and technical issues are aligned from the first moments of the assets' operational lives, then he is already in compliance with the ethos of the Asset Management System as defined by the ISO 55000. This vision is therefore undoubtedly a major lever in the implementation of a functional and holistic organization. In my experience as an ISO certifier, I'm often brought to visit industrial plants or infrastructure networks where it is seemingly impossible to produce an exhaustive Asset Register. Yet it is most likely that if I were to discuss this detail with members of the board of directors of one the organizations operating these plants, they would be outraged; this is symptomatic of the level of intersegmental disconnection at hand.

It therefore strikes us as obvious that it is essential to standardize the ways in which the different actors perceive the installation function. However, such a transformation in representations can only emerge from a corporate will. Management and leadership must therefore become aware that the failing and perpetuating scheme at work is primarily the result of the pressure that they themselves apply on their employees. Calls for quicker and more effective processes of innovation, production, or profitability are continuously hammered on the workers by decision makers and, in the end, harm the quality of the installation.

The decision makers themselves must therefore realize that an optimal and completed installation is a precious asset, both in terms of anticipating future counter performances and in terms of provisioning the penalties that such counter performances would induce. Indeed, in the few cases where this period of installation is optimally considered, one can see how easy it is to begin so early to calibrate a potent Enterprise Resource Planning (ERP) that will give substance to all of the important information and therefore contribute to giving shape to the Asset Management System. Results are visible as soon as management becomes aware of this issue's critical importance. On the one hand, of course, this approach involves mobilizing more resources, and is usually more time-consuming—but the gains realized are considerable. The period of installation is also most adequate regarding the framing of the operation's maturity level and the calibrating of the maintenance that one seeks to implement.

If these aspects are not assessed quickly, it is indeed likely that operations will begin "in the dark" (regarding crucial information), that operators will be hired on the spot and asked to begin work without the proper formation, and that repairs will be made on the machines without anyone able to boast a particular knowledge of the assets' specificities or of the means that would make them truly performant. Overall, the situation will be that of a critically low level of maturity. In this type of situation, the absence of a pre-established strategy results in making operators behave like simple button pushers and maintenance agents like pure parts exchangers. On the contrary, a proper initial calibration allows for the necessary observations to be made in order for a targeted organization to be implemented so as to live up to the pre-established level of maturity. Thus, one is able to know very early on in the operational life the quantity and the nature of the resources that must be mobilized and of the capacities that must be deployed, as well as the form of strategy that must be adopted.

It is common knowledge that Asset Management, as well as the quality processes that have flourished across Europe for the past two decades (via the European Foundation for Quality Management), place these maturity scales at the core of their approaches seeking to enhance the global management of organizations in general, and industrial installations more specifically. However, in order to engage itself on the Asset Management journey, an organization must first proceed to a gap analysis in order to assess the transition that must be undertaken in terms of maturity. The notion of "gap" should not be understood as a pejorative statement, but rather as the essential starting point of the Asset Management processes to be implemented.

4.3 Reliability: The DNA of Industrial Assets

It is urgent to establish the idea that the installation process has a much more crucial impact on the assets' future reliabilities than we would intuitively imagine, in spite of all that one can do to determine these reliabilities in the project phase. It is indeed during the installation period that one can provide guarantees to the operational reliability standards that one has already measured. In order to shine a light on this issue, one can resort to using the example of redundancies.

Whether in the installation phase or later on in the life cycle, it is essential that the teams tasked with Operational and Maintenance realize that the constructive character of equipment fitted with redundancies must be understood for what it is. A redundancy is indeed not designed to endow the Operational and Maintenance sphere with the illusion of comfort. This idea is based on a misconception, or a bad habit, developed in the industrial world. The actual value of a redundancy lies in its own nature as a solution that favors the accomplishment of the assigned objectives of the asset's operation. One should thus be careful to not misunderstand the finality of the redundancy at the time of the development of Operational and Maintenance strategies. It would moreover be very dangerous to consider that when one sets two pumps to run in parallel so that when one is active and the other on standby, one has succeeded in creating an ideal situation for practicing a preventive maintenance operation of the first pump. At the core of this danger is a true conceptual problem. Indeed, in an analogical situation, this would be the equivalent of taking advantage of the fact that an ER patient is ventilated by two oxygen bottles (one active, the other in standby mode) to proceed to an action of preventive maintenance on the paused oxygen bottle.

This rationale—which consists of stating that since there is a redundancy, we can allow ourselves to unplug one of the machines to proceed to its maintenance—is erratic. The redundant bottle is not here to replace the other in the event of a maintenance operation—it is here to save the patient's life in the event of a failure! And the same could analogically be said for any redundant asset in an industrial plant. As we've said, the installation phase is an ideal period to ensure that one is set to stick to the agreed-upon levels of reliability. In the same line of thought, the installation phase should also mark the beginning of the assessment of the aspects tied to maintainability—the aptitude of assets to be maintained or repaired. Indeed, even though maintainability should theoretically have been calculated as early as the design phase, it would be regrettable not to take advantage of the installation phase to validate *in situ* the assembly and dismantling times that will serve as a base for calculating the operational maintainability.

In light of all these elements, I would like to offer an observation: not taking advantage of the opportunities of the installation phase will inherently make the operational life less adapted to the evolution of its mission profile, which means that it will inherently be less flexible in terms of responding to the evolutions of demand, production, or technology, or to the modifications that will have to be made on the assets throughout their life cycles. Yet, we are supposed to build plants with a long-term vision, and anticipate the fact that a plant is usually not a fixed entity but an apparatus that can be changed according to the specific needs or the technological improvements of the moment!

In this perspective, one can easily understand the importance of using this "anticipatory" phase to determine the kind of information platform that will be used over the course of the operational life. Of course, we are well aware that new tools (especially in terms of soft and hardware) will inevitably be brought into usage throughout the assets' life cycles—this is a permanent factor in the context of continuous technological progress. This is all the more reason for the organization to engage in a proactive attitude and to successfully project itself on the long term in order to avoid making sub-optimal decisions in the operational phase. Very pragmatically, we know that it would be naïve to expect operators to restrain themselves from using innovative and efficient tools simply because they were not included in the initial plan. One must therefore preventively assume that the operating teams and employees will exceed the frame of an initial masterplan. Furthermore, a perimeter of industrial assets should also be designed in order to be resilient to the evolutions which will inevitably occur.

In conclusion, what can we derive from these observations? Primarily, that Asset Management now constitutes a very positive lever in the mitigation of the contrast that still prevails in the installation phase between two seemingly contradictory corporate desires: to operate *quickly* and *at the lesser cost*. Yet, the points discussed above are unanimous in showing how many wasted opportunities are encountered in the industrial world in terms of the capacities of Assets/Organization pairs for the remainder of the life cycle that is being constructed and installed. Indeed, we know that to engage in construction at the minimal objective price means taking the risk of generating enormous expenses in order to allow the assets to adapt to the evolutions of the market, in terms of debottlenecking (the augmentation of their productive capacity) or of rescheduling of their end-of-life (in order to delay the mobilization of CAPEX, which will be provided by investors and shareholders). In other words, if an organization chooses to resort to a life-cycle-costing perspective, the amount of money deployed will remain the same, but its return on investment will vary greatly. By mobilizing this sum in the cash flow with present or future values, one is perfectly able to optimize the value extraction of assets in a global vision of costs, value, and performance as early as in the installation phase.

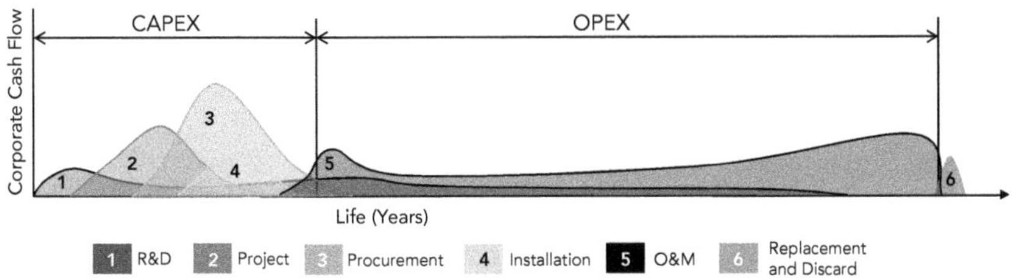

FIGURE 4.1 Value can only be extracted during the asset's life cycle.

Furthermore, it is clear that to concentrate the pre-evoked assignments on the installation phase generates indisputable profit. The installation phase constitutes a major reservoir for future value extraction, if only through the opportunities that it provides for mitigating or resolving the imperfections inherited from the project phase, in addition to being the ideal moment for preparing the operational phase. And indeed, establishing all Maintenance and Operational strategies in this preliminary phase will have a facilitating effect on the commissioning stage. On top of all this, this process of risk awareness and anticipation over costs and performances constitutes an important factor in avoiding setbacks later on in the life cycle. It is therefore a necessary and efficient approach to the construction of the processes that will be the base of the Asset Management System of an operation ready to embrace greater success.

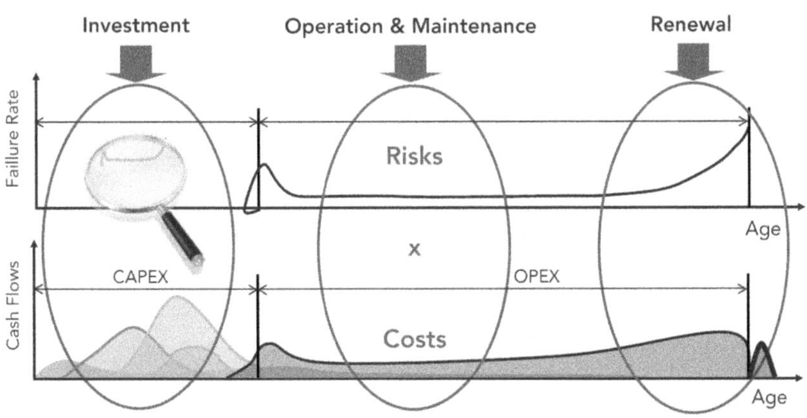

FIGURE 4.2 The relationship between risks and costs at every step of the life cycle.

CHAPTER 5

From the Commissioning of Assets to Their Operational Maintenance

The commissioning of assets is a phase intended as the setting in motion of industrial installations and infrastructures, and often aims at the realization of a number of precise objectives, both global and specific. It seems obvious that in the perspective of the successful commissioning of an industrial installation, one of the priorities should be the resolution of the interface between the installed equipment's manufacturer and the team tasked with the installation on the one hand, and the managers of the organization that will come into possession of the assets for (essentially) their useful life on the other. Indeed, for a number of reasons, these two groups display two potentially irreconcilable conceptions of the machines. In one of these conceptions, assets are regarded as deliverables, while in the other, they are seen as the answers to the requirements exposed in the design brief. The preoperational phase, which can teach operators a great deal about the machines themselves and more specifically about the Asset Systems, facilitates the coincidence between these two distinct representations and helps them to target common objectives; yet, it is quite likely that none amongst our readers could assert that they'd ever witnessed a situation in which the expectations of operators and the manufacturer's deliverable were rigorously identical.

5.1 Commissioning and Maintenance: Between Two Worlds

To shine a light on this somewhat equivocal situation, one should recall that the synchronization of requirements and realities has historically depended on the actions of operational managers. If we were to study this transitory phase through the prism of Asset Management concepts, we would probably get the same results, but at a much greater amplitude. One should understand that any action of commissioning is practiced in a context in which the industrial investor wishes to see his installation become operational and apt to realize the products that it was designed to manufacture as soon as possible. The commissioning is therefore largely a *context of emergency*.

Hence, it should come as no surprise that the commissioning stage comes with its share of discoveries. Some amongst these discoveries are usually expected, since installations are increasingly reliant on new technology, which has not circulated long enough for experts to gather sufficient feedback to proceed to clear anticipations over its performance. Unfortunately, the vast majority of these discoveries come as total surprises. Our goal here isn't to criticize the fact that these "surprises" occur, but rather to show that these events have happened repeatedly and in a structural manner for decades while organizations have seemingly not learned to anticipate or to handle these setbacks. Without going into tedious details, one could very generally say that in this domain we are often faced with very unpleasant surprises, which can for example take the form of counter-performances from the equipment or the organizational structure, or be tied to neglected aspects of the installation. Many of these setbacks could have been anticipated but have not been solved in due time. In my opinion, the most serious of these issues is that of staff level, both in terms of manufacturing and of operational management. Indeed, it is frequently observed that a considerable share of the workforce has not been properly trained to deal with the situations that they are faced with, even if those situations are typical of the commissioning phase.

The surprise in this case is not due to the work realized by these teams who struggle for weeks on end to proceed to the calibration and the parameterization of complex installations. It is rather due to the organizational dysfunctions and to the setbacks inherent in technical improvisations and that are obviously partially blamable on the teams' lack of expertise, but more rightly so on the hastiness that is asked of the workers in the context of the *ramp up* phase.

5.2 The Industrial Diapason

In this stage of the asset's life cycle, Asset Management provides a great opportunity to optimize the anticipation of gains through the running-in of the machines and the further development of the workers' understanding of the installation's physiology. Indeed, those factors have historically allowed organizations to validate or, better still, to calibrate the opportunities of potential value. This process has recently been facilitated by the emergence of innovative and performant tools and methods in the industrial world. One should also keep in mind that commissioning is very important to Operational and Maintenance managers, since they can use this stage to study the strengths and weaknesses of the assets as well as the specificities of the tools they will need to deploy when functional imbalances between risks, costs, and opportunities arise. This aspect brings more practical sense to the notion of *trade-off*, which is central to our culture. By means of an analogy, why should one bother to play an instrument if you haven't properly tuned it beforehand?

It is the *proactive* effort (which consists of using the time of commissioning to get familiar with the installations' specificities) that makes the form of alignment promoted by Asset Management so innovative in terms of elaborating a new common mindset shared by operational managers and the board of directors. This new approach seeks to avoid or mitigate factors of vulnerability that would have an important negative impact on the remainder of the asset's life cycle by increasing the control over certain details of the installation's operation and more specifically over the risks, whatever shape they may take. Indeed, dysfunctions that are already present in the early stages of the assets' useful lives can turn out to be catastrophic in the long run. Imagine, for example, that one were to shoot an arrow towards a target from a distance of 100 meters (the target, of course, is a symbol for the assets' operational life cycles). A few degrees of miscalculation could send the arrow flying a full meter off target! In the same way, a tiny dysfunction at the beginning of the asset's life cycle can have a crucial impact on its global performance.

A number of my colleagues in the field of Asset Management are even reflecting on the idea of anticipating the nature of the trade-offs and of the main "critical" assets by preventively preparing the corresponding treatment. In other words, this implies asking, very early on in the process, for each asset and for the global installation, the following question: what kind of functional imbalances (between cost, risk, opportunities, spare parts, degradation, inspection, etc.) can we anticipate, rather than eventually be faced with later on in the life cycle?

What this implies is that commissioning is simultaneously the best and the worst moment to become aware of deficiencies (in the workforce); far from being restricted to the strictly technical skills, these deficiencies can also be tied to the staff's appropriation (or lack thereof) of the corporate strategy. In other words, it is crucial to understand that the commissioning phase is an ideal time to ensure that operational managers have come to terms with the scope of their task as *business engineers*. Hence, the engineer must learn to outgrow his technical task and understand the contribution of his own input to the realization of the challenge set by his organization's business objectives. Therefore, it could be assessed that an efficient business engineer is characterized by his understanding that the function, which consists of treating the parts and functionalities of the industrial installation, must be exercised in the frame of an accomplished alignment with the expectations and requirements of the corresponding business.

In this perspective, we could name a few other important and yet often neglected elements of the future Asset Management System that will underpin the rest of the assets' life cycles in order for these assets to generate gains but also and more importantly to endow the organizational values (pecuniary or not) with an increased consistence. Thus, we could evoke the acuteness with which the organization responsible for operating the installation will handle the physical degradation of assets as well as the methods that it will have to implement in order to inscribe their performances in the long haul. Let us also keep in mind that every piece of equipment comes with its own inevitable degradation kinetics, which, sooner or later, will bring about the failure of its function. A real mastery of methods that go beyond purely technical practices is therefore necessary in order to ensure that the balances reached between risk, cost, and performance remain aligned with the expected life cycles of the assets.

On this point also: what a shame it would be to pass on such an opportunity to begin the processes of calibrating the performance indicators (at the very least in technical and economic terms) and to distinguish between indicators tied with the *burn-in* (the first stage of the life cycle, subject to the infant mortality of assets) and indicators that correspond to the operational performance of the machines!

It should be observed that this distinction is anything but a perfectionist's whim; it actually coincides with the transition operated in the commissioning phase between the budgets allocated to the ending project phase and those that will serve in the freshly begun operational phase. Therefore, one can understand that the commissioning of assets truly represents a transitory segment in the machines' life cycles, whatever the surprises and setbacks that may arise from this stage. Once it is over, despite all of the hindrances and obstacles that may have emerged, the installations will start working, and they will start producing.

In order to close this chapter on a more positive tone, one should insist on the hope that the emergence and democratization of Asset Management practices represents in this field. Indeed, the multiplication and the growing availability of adequate tools (which act as so many industrial diapasons) as well as the consistent increase in the maturity of contractual engagements could well allow for the resurgence, in the near future, of theses as brilliant as those of Mr. Duane's "reliability growth," which would be focused, in our discipline, on the notion of "value growth."

PART III

The Operational Life

If the preoperational phase allows one to assemble the conditions that will favor an optimal extraction of value in the operational phase, the processes deployed in this preliminary stage do not directly have a hand in the extraction itself.

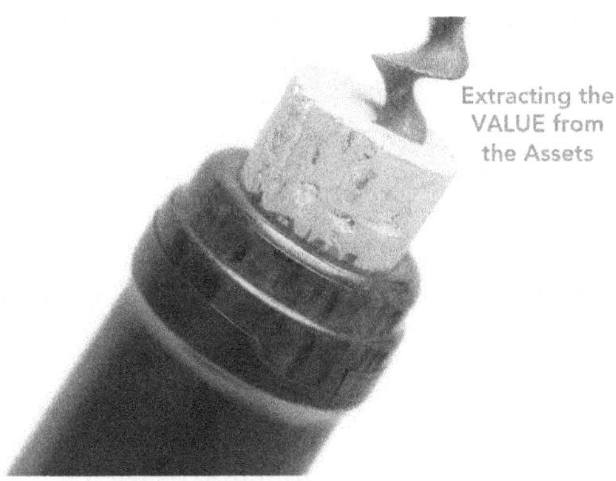

Analogy illustrating the difference between Asset Management and maintenance

Similarly, one should keep in mind that efficient maintenance work by itself cannot suffice to extract value from an asset either, or at least not in the optimal sense that Asset Management advocates. To shine a light on the limitations of a maintenance strategy realized without recourse to a global Asset Management perspective, one could resort to using a seemingly trivial analogy: that of the conservation of a bottle of wine.

Where do we place the value of a bottle of wine? Unless we are dealing with a collector who might consider the bottle itself to be endowed with a proper value, we can agree that it is the wine that confers value to the item. Thus, the period over which the bottle is stored in a cellar is the equivalent of the maintenance period of this asset. The asset is tended to and preserved through the combination of different factors (in this case, temperature and humidity) but it is only once the bottle has been uncorked and the wine is being served that one gains access to the value acquired by the realized investment. It is then that the customer's satisfaction may be expressed and that he truly proceeds to the actual extraction of its value. Hence, we could assert that maintenance sets the stage for the extraction of value—it is therefore a necessary component of Asset Management, but it does not suffice in itself to reap the benefits generated by the assets. Maintenance is an important vector of return on investment, but it may by no means generate this ROI on its own, nor does it deliver any direct service to the consumer (the main element in the recognition of the created value). As we know, it is the sum of all these stakeholders that makes up the value that Asset Managers seek to extract from the machines.

CHAPTER 6

What Is the Operational Life?

6.1 Maintenance and Asset Management: A Necessary Partnership

As soon as Asset Management became regarded as a serious discipline providing precious input, the first confusion about the object of this new discipline and the specific object of the maintenance industry came to light. Whether this tension has emerged from a certain jealousy or from a sincere desire to do right is ultimately inconsequential. All that matters are the facts. And they are indisputable: those two fields are undoubtedly distinct, even though intellectual integrity forbids us to oppose them—quite the contrary. Those two approaches are bonded by a form of necessary partnership.

In order to be quite understood, I will allow myself to open this presentation with its conclusion. One can undeniably perform high-standard maintenance without relying on Asset Management. However, it is unthinkable to practice Asset Management without already possessing a perfectly functional maintenance strategy and practice. We will now come back to my "wine bottle" analogy in order to demonstrate this phenomenon.

A wine bottle is a glass object, equipped with a cork, a label, and a bottleneck seal. These are the physical elements that make up the said bottle. With the exception of collectors, the people who buy wine are more motivated by the contents of the bottle than by the object itself. In other words, the investment is intended to bring pleasure, which will come at the time of consumption and which can be described, in our vernacular, as the extraction of value from the object. In short, the finality of maintenance is to preserve the objects that contain the value that an organization will then strive to extract. Taking care of the bottle, which is equal to maintaining the object that contains value, is therefore an essential activity that must be both effective and efficient in order to provide the optimal conditions of value extraction in due time.

How can one pinpoint this propitious moment for the extraction of value? If the object is a bottle, then intuition will tell us that the ideal moment would be at mealtime. But in the perspective of industrial activity, these two moments of the life cycle (maintenance and value extraction) are almost simultaneous. Unlike for the bottle, which undergoes two distinct cycle segments (conservation and consumption), these two activities are often coincident in the case of an industrial asset. Our analogy therefore fails to clearly highlight the reality of these processes, which coincide in the world of infrastructures and industry. However, one should keep in mind that "simultaneous" does not mean "identical." Why, then, do the temple-keepers of the world of maintenance relentlessly stir up controversy around the idea that this new discipline would, in the end, be a mere dispensable graft to theirs, while it is clear that these two sectors are set in distinct temporalities and serve clearly different purposes?

Because all too often, they've witnessed the emergence of Asset Management "around" their activity, since Asset Managers regard maintenance as an essential factor for success.

Because they've long been deprived, as I once was, of the possibility of being acknowledged as contributing factors to the extraction of the value of the assets they were preserving.

And finally, because the concentration, in this sector, of professionals set in the technical world (and notably of engineers) has naturally had a restrictive impact on the maintenance field, by making it a practice of technicity, operated by technicians and for technicians.

Our task is therefore to demonstrate, through the use of conceptual and objectives elements, that there is no valuable reason to oppose the two disciplines, which we should on the contrary attempt to reunite on the practical plan.

In order to further develop the subject, let us begin by a quick definition point:

An industrial asset, as defined by the ISO 55000, is an item (or entity) that represents a potential value to an organization, in the sense that the organization expects to extract value from the item. This definition is of an invaluable precision. The aim, for the operator, is not solely to preserve or to maintain the asset indefinitely, but indeed to extract a form of profit. The task of the Asset Manager is therefore clearly distinct from that of the maintenance operator; indeed, we know that maintenance seeks to ensure that machines are available, or, to put it differently, functional. In other words, it allows for the optimization of the conditions in which the operator will proceed to the extraction of value, but it does not participate directly in this extraction.

However, availability is the cross product of the intrinsic reliability (the ability not to undergo a shortage) by the maintainability (the ability to undergo repair in the event of a shortage). Since reliability is an attribute handed over to an asset by its manufacturer, the maintenance function may only take notice of its level (good or bad) at the time of acquisition. The maintenance function is therefore quite incapable of seriously impacting on the level of reliability, regardless of the frequent bragging perpetrated by a number of colleagues who claim to maximize their assets' reliability levels when in truth they only control one among many levers that shape this reliability. In the best cases, maintenance operators *tinker*, or they behave as expert mechanics, with all of the limitations inherent to this function. In this regard, I often point out that we (along with doctors, who act, after all, as the maintenance function for the human life) are unable to alter a DNA profile after birth.

As for maintainability, it falls well within the scope of the maintenance function since even though the ability to repair an asset is inherently attributed by its manufacturer, it will largely be conditioned by the physical, functional, and managerial organization of the asset's operator.

Generating availability is therefore the central function of the maintenance department, and yet in its drive to target a truly profitable availability, the technical manifestations of maintenance are unfortunately insufficient.

Everything has changed in our ways of life and our work habits; in the level of requirements; in the quality criteria of products and services; in the monetary world; in the very tools of productions; in the behavior of children (the "Y"

Generation) in particular; in the face of a digitalized world that seems within reach; in the managerial modes of our superiors, etc.

Hence, when a number of maintenance agents assert that "they've been doing things this way for 20 years and they can't see why they should reconsider anything" when we question them about an anachronistic vision or a number of outdated maintenance practices, we are clearly entitled to wonder whether they alone are impervious to change.

What transpires is that Asset Management may not be defined by an operational character: it rather finds its basis in a strategic manufacturing, bent on the quest for a balance between costs, risks, and opportunities—the infamous *trade-offs*. On the contrary, maintenance is characterized by an operational activity, that of preserving functioning assets.

6.2 Life Cycle and Alignment

In order to bring an end to this fruitless controversy, we must insist on the perspective focused on the assets' life cycles, which is fundamentally inscribed in the program promoted by Asset Managers and widely absent from reflections revolving around pure maintenance (see Figure 6.1).

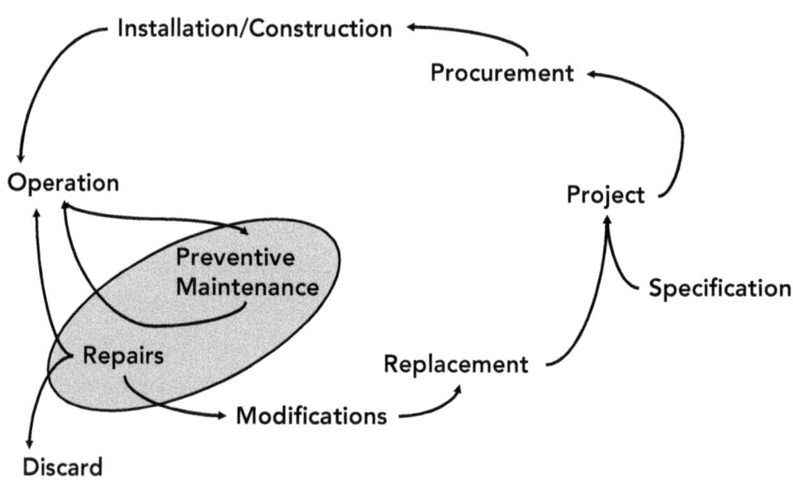

FIGURE 6.1 Segments of the asset's life cycle

Indeed, in our discipline's specific lexicon, *asset life cycle* refers to the continuum that begins with specification and covers the whole of the asset's operational life and end-of-life. Maintenance plays its part in this process, but it remains a segment among others. It is therefore clear that one can practice maintenance without practicing Asset Management, but that the reciprocal is invalid.

In order to shine a light on the meanings of this figure, we will resort to using another analogy. Let us consider the case of a football team. It is an asset, composed of 11 segments (the players), or in other words, of 11 periods that, combined, make up its life cycle. Let us imagine that the team in question is outstanding in that it features, for every position, the best possible candidate—an "All-Star" team where each function would be realized in an optimal fashion. That was more or less the situation of the Brazilian team on the eve of the 2014 World Cup, which had at its disposal, if not *the* 11 best players in the world, at least a set of players among which each one was individually known to be one of the most performant in the world for his given position. This All-Star team, which boasted exemplary "individual positions" (the segments), was however defeated 7 to 1 by the German Mannschaft, in a memorable and indisputable game. Other than the obvious merit of the German players, what may have caused the Brazilian team to fail so totally to withstand such a challenge?

The answer is truly quite simple. It is through the elaboration of a global strategy, a collective process and a total coordination between the players that Germany was able to implement a winning tactic. Conceptually, what happened that day is linked to a phenomenon well known by the adepts of the theory of constraints. Indeed, the works of Descartes, Goldratt, or Bellman demonstrate that the sum of the optimums is systematically different than the optimum of the sum. In other words, bringing together a set of individually optimal segments does not suffice to guarantee the success of the implemented system.

It can easily be observed, to support this argument, that contemporary organizations are quasi-systematically structured as "rakes." "Rake-shaped organizations" qualify the type of business-model where a CEO or president is placed "at the top" of the model, with underneath him either one or plural rakes each supervised by their own manager. Typically, the distinction operated between these various parallel managers is either a functional or a geographic one, and in an Asset Management perspective it would make as much sense to represent such an organization by displaying in parallel the managers of the life cycle segments in a logical sequence ranging from the design phase of assets to their timely dismantlement.

This very common type of structure brings about a type of functioning where every year, the objectives established at corporate level are allocated to each "rake" manager. What this implies is that by the end of the year, when results are to be measured and assessed, the sum of the objectives allocated to every one of the life cycle segment managers should amount to a list of results corresponding or nearly corresponding to the targeted results. Unfortunately, *this would be too good to be true*, and there are many reasons to explain why this extremely common approach does not generate success. Indeed, in practice, this method does not operate exactly as it "should;" hence, the analogy provided by our discussion of the Brazilian football team above. Not only is collective coordination much more efficient than cumulative individual performances, but worse: it is frequently observed in the context of organizational functioning that success in fulfilling the objectives of one or plural segments of the rake is achieved to the detriment of the success of one or plural other segments.

This sets a great challenge for leadership in general, and for Human Resources managers in particular, since they are accountable for the evolution of careers and wages. The segmentation of yearly objectives is definitely an argument in favor of measuring performance at an individual managerial level, even though in order to avoid the stated pitfalls it is clear that Human Resources should increasingly take into account the result of the holistic and collective dimension of the produced labor, without neglecting of course the importance of individual performance. Such a tendency would allow for an increase in overall efficiency as well as avoiding the more perverse effects of the inevitable "rake."

Let us come back to Figure 6.1: since we see that maintenance is represented as two segments, it is clear that it may by no means suffice to ensure the success of the management of an asset on the whole of its life cycle—even less so if the maintenance is not operated in the context of an operational deployment aligned with the other elements of this life cycle. Asset Management is thus designed to act as a holistic observer facilitating the elaboration of a strategy for collaboration and sharing between different agents of the assets' life cycles.

These reflections bring us back to an anecdote that I've previously referred to in the foreword to this book. During my time working as a maintenance manager in the Brazilian public transportation system, I was the first among my colleagues to question the principles of functional analyses in order to achieve a transition from the "object" vision of part-by-part maintenance to an approach more bent on the maintenance of functions.

At the time, we were only beginning to experiment with the RCM method (Reliability-Centered Maintenance), which became in its time a true stepping stone for a more holistic approach to a function that has historically been associated with technical procedures. I soon realized that each segment of the life cycle was endowed with a designated manager within the organization. I myself was tasked with supervising maintenance. Each of these managers was handed annual objectives and, since we were competent, these objectives were usually fulfilled. Hence, we were in a situation where we'd seemingly realized the "sum of the optimums"—but none seemed to question the manner in which we obtained such results.

I began to realize that the procurement division, which was tasked with the optimization of the acquisition of assets and spare parts (both in capital and operational expenditures), had become accustomed to acquiring cost-effective but technically low-quality equipment, without anticipating the fact that the savings in terms of procurement would invariably have a negative impact on the maintenance budget for years to come. By seeking at all costs to mitigate CAPEX, the buyers were therefore playing a clear part in increasing our OPEX, which in turn would limit the budget allocated to my maintenance department for the following years. It is important to grasp that the costs we are discussing occur in the long run, as OPEX can emerge until the very last stages of the assets' life cycles.

The fact that I mention this practice does not imply in any way that I should wish to blame procurement managers for the process, since they were merely following the expectations and demands of their superiors. To this day, I remain certain that the buyers were acting in good faith. The problem on which I wish to shine a light is not an operational issue, but rather a systemic problem that derives from the constraints imposed by top management.

Indeed, corporate managers are only playing their part when perpetually demanding more and more from maintenance managers: the latter are asked to mitigate costs, to optimize efficiency—in short, to improve their results while reducing their expenditures and the means at their disposal. Indeed, in spite of themselves, they are often complicit in holding back the evolution of the emerging methods and processes of the maintenance function. When considering such a task, one can only be reminded of the famous Einstein aphorism that states that insanity "is doing the same thing over and over again and expecting different results."

Asset Management is precisely the alternative path through which the sought-after result may be obtained, since it encourages a modification of the very structures of labor and operation, in rupture with a model whose limitations are well understood at this point.

We therefore intend to work differently, using a structured decisional approach and a clear and verifiable process taking into account the assessment of risk, that of the financial and non-financial objectives (particularly in issues linked with security or the environment), and of the short- and long-term consequences of the decisions made. Our goal is to practice a maintenance informed by technical and managerial progress, which may facilitate the optimal extraction of value from assets. Furthermore, we are now aware that the best indicator of the extraction of value from an asset over its life cycle is minimal *life cycle costing.*

6.3 I Leap One...

To illustrate the need for a structured decisional approach, conflating maintenance and Asset Management, let us imagine what a talking machine would say to substantiate this argument:

> *"Where the hell have you been? You should have visited me long ago. Ever since you set me up in this plant, you came to take care of me once a year, without fail. You came to check my panels, replace my parts, oil them, I was impressed. You even went above and beyond, carrying out unnecessary operations, which I didn't enjoy, but hey ... at least, you showed you care for me.*
>
> *"Now, I can't understand it. How could your considerate self change so radically overnight? The rumor is that you will visit only once every eigh-*

teen months, is it true? I can't believe it. While my engine and my pulleys can do without preventive maintenance for a year and a half, my belts certainly can't! If they break and fly across the workshop, there better be nobody standing close by.

"I bet you want to save money on me. Since you have logical and practical sense, you have tried to gather all the preventive maintenance tasks in a single sitting, and not bother the rest of the time, is that it? This way, you don't hamper production, which hates it when I'm down. This must be it! I can also guess what will happen in eighteen months. You will put on a big show to make up for everything you've foregone in that time. You will strip me of all my parts, and place them back on, which is very painful for me. Don't you realize that it weakens my reliability?

"After all these years together, I expected you to know me and my needs better. If other machines can make it through that period without maintenance, good for them. But I can't! What rationality led you to decide to treat me so?

"If you think you'll save money on my back that way, you are terribly wrong. I am going to become much harder work. And don't come crying later, it's your fault, not mine. I can't work miracles."

If machines could talk, they would question the criteria upon which industrial companies modulate the frequency of preventive maintenance when they need to cut costs.

Every maintenance manager knows what to expect at the end of the year, when the company draws its budgets for the following year. The question isn't whether the boss will ask them to cut preventive maintenance costs, or even when—it is by how much. The same goes for every industrial company in the world: steadily, every year, maintenance costs, along with a few other departments, are trimmed to increase competitiveness.

This situation has increased in frequency since the 1970s, following the first oil shock, and more recently with globalization. To illustrate what has been happening, let us make an analogy with a spring. Imagine a spring at rest, fully loosened: this was the situation of maintenance costs 30 years ago. If the spring is tightened by 10%, the required effort is a small one. In fact, with an "E"-strength energy, we easily manage to tighten it by 10%. However, the spring's elasticity coefficient is "k," which means that the more it is tightened, the stiffer it becomes, and the harder it becomes to tighten it. So, we started out with an "E" energy to tighten the spring by 10%; but, to tighten it by a further 10%, the energy required will necessarily be greater than "E."

Let us now imagine the state of the spring after it has been relentlessly tightened for years. In some companies, the spring has reached a state where it can be tightened no further, yet management continues to require cuts in maintenance costs. Why wouldn't they, since maintenance managers have always proved able to tighten the spring a little further? Besides, cutting costs is always useful, and everything in life is perfectible. Yet, beyond a certain point, the energy required to reduce costs is greater than the resulting gain. Since we don't take into account the basic fact that cutting costs increases our risk exposure, and since we are bound to follow orders, we keep on tightening costs, and thereby introduce an imbalance between the energy required for tightening (with its resulting risk exposure) and the expected benefit.

At any rate, this does not mean that we should stop reducing costs. In most companies, significant margin squeezes are still possible and, even better, with little effort and without increasing risk exposure. In many cases, both can be achieved at once: reducing both costs and risks. Achieving this requires changing our ways of reducing costs and of implementing decision-making. This is what we will discuss now.

For the maintenance manager, increasing the periodicity of machine maintenance is an efficient way of reducing the costs of his department. Consider a maintenance operation that is carried out every 24 months at a cost of 6000 euros: spread over time, the expense is 250 euros a month, right? If the frequency of the operation drops to every 36 months, the expense is brought down to 166 euros a month, thereby saving over 30%.

Undeniably, modulating the periodicity of preventive maintenance tasks is a fathomable logic that meets the budgetary objectives. The maintenance manager goes over his board and tries to decide where to intervene: "will it be the can of the compressor or of the bridge crane? The belt or the spur gear?" His experience tells him where the margins are and where the periodicity can possibly be increased. Nevertheless, his decisions are influenced by some variances that might undermine their efficiency, one of which is the loving care. "I won't operate on this machine because it is too special. I will not carry out this task because it is inseparable from this other task...." He will tend to leave some machines untouched, and to keep repeating the tasks that he likes best.

Another bias in the maintenance manager's attitude is evidenced in the way he calculates the modulations of the periodicity of tasks. He uses multiples to calculate new intervals in preventive maintenance. For example, a task that is carried out every 6 months will be postponed to every 9 or 12 months; a task carried out every 10 weeks will be postponed to every 15, 20, or 30 weeks. This process is both logical and simple: there are hundreds or thousands of maintenance tasks to complete in the plant, and every one of them must be mathematically planned, to optimize the technicians' worktime and the machines' productivity. Therefore, it is easier to proceed with "maintenance while we're at it"—which, by the way, is another significant bias.

Consider a machine that has its mechanical components serviced every 12 months and its electrical components serviced every 18 months. When modulating the preventive maintenance plan for this machine, the maintenance manager will reason in the following way: "Since Dubois is scheduled to check the electrical components every 18 months, I might as well shift Dupont from 12 to 18 months. That way, they will check the electrical and the mechanical in one go, replace the belts and the pulleys, lubricate, all at the same time." Do you get what "maintenance while we're at it" means? "We have to stop the machine for the electrical maintenance anyway, so while we're at it, let us also squeeze in the mechanical maintenance."

While the maintenance manager's reasoning is pragmatic, it does not account for the risks involved in increasing the interval. The pulley might be able to go 18 months without service, but the belt might not. The same goes for other parts that are scheduled to be checked along with the belt. Reducing the interval below 18 months would most likely reduce the risks, but how can we sustain this line of reasoning if we have no way of accounting for them? Furthermore, how do we calculate a new frequency without reverting to a simple multiplication?

Let us now consider a practical example. Forced to cut costs, the maintenance manager of a construction company has decided to modulate the frequency of preventive maintenance for a shield tunnel, service for which runs up a bill of $24,000 a month. With the classical method of multiples, he has increased the interval from five to ten months, and thereby brought the cost down to $12,000 a month (Figure 6.2). Great news, isn't it? Logical and simple. I can picture him going over his board, noticing that this specific maintenance task is penned in every five months, and thinking: "I only have to postpone one of these tasks, keep the one that occurs every ten months, to cut costs by half." Yet, by consistently thinking in the "postponement" mode, we expose the machine to considerable breakdown risks. Take a look at Figure 6.3: the overall cost of this maintenance decision (that reverberates beyond maintenance) in a ten-month period is double the cost on a five-month period!

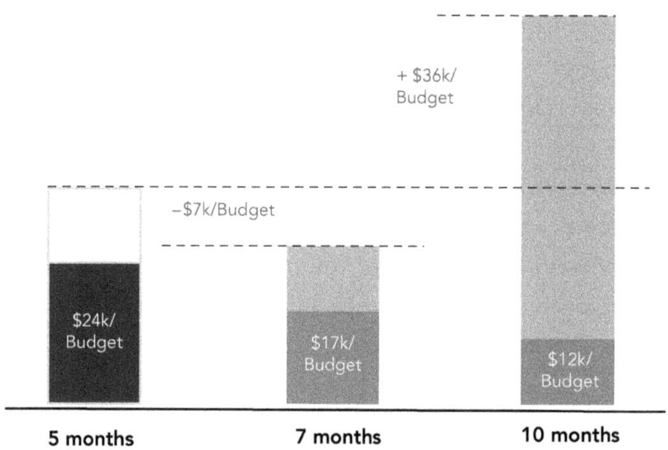

FIGURE 6.2 "Leap one..." versus optimizing the risk/cost ratio of maintenance intervals

The risk of failures, which already showed on some parts after five months, will obviously have doubled after ten months. Furthermore, new risks have appeared on parts that couldn't handle the increase of the interval. This was evidenced by running a simulation of the risk/cost relationship of modulating the maintenance periodicity. This simulation also revealed the optimal periodicity: seven months, a number that would never have been found with the classic "multiplying and postponing" method.

This is not a one-time occurrence. As a consultant, I have run hundreds of similar simulations for companies looking to cut costs by increasing the intervals of preventive maintenance. Approximately 70% of such increases that we have analyzed ended up costing more than the initial savings, precisely because they were implemented without accounting for the risks derived from modulating the frequency.

In the face of these case studies I mention, one might legitimately wonder why maintenance managers refuse to consider, over a given time period, non-multiple frequencies, if only to uncover these alternative results. The fact is that they firmly refuse because they don't dare to imagine the mess this will induce in their schedule. Try to imagine the daunting task it would be to conciliate round number–frequencies with frequencies that aren't so: they would have to reorganize hundreds of tasks, along with thousands of related logistical tasks, and this would seriously increase their workload. Strangely enough though, maintenance managers do not forego such reorganizing when it is for other reasons. This happens often in a plant; for example, when a new machine is implemented, when production rates are modified, or when heavy corrective maintenance must be carried out without delay—which tends to be on a Friday night or at 5 o'clock on a Sunday morning. The problem really lies elsewhere: it is the desire to optimize downtimes by regrouping various tasks around a main task, be it for convenience, for personal preference (usually subjective), or by chance. Unless the maintenance manager is a brilliant poker player (and even then), he will have little to no chance of obtaining positive results in terms of costs and risks by proceeding this way.

There is another (valid) reason for calculating frequencies with multiples: they limit the conflicts that naturally occur with production. General management puts production under huge pressure to ensure that production never stops or slows down, and, naturally, they resent maintenance's ability to stop a machine whenever it feels like it. Calculating preventive maintenance intervals with multiples, as scarcely as possible, and according to the "maintenance while we're at it" principle is therefore in the interest of production: we shouldn't change anything, which is legitimate considering the constraints that are imposed on us. If he voiced the desire to change this way of doing things, the maintenance manager would without a doubt alienate the whole company. This includes general management, which usually sides with production, because both would like machines to run without pause. The balance of power in industry is akin to a dice where five sides read "production" and only one side reads "maintenance." There is a reason why "maintaining" oneself in the game is so tough! What we are up against is a triple spiral of failure. This consists of the maintenance manager spiral that sticks with multiple numbers on the back of the belief that he is optimizing his work; the head of production spiral, who would like machines to never stop running, as if this were possible; and the general director, who doesn't realize that he will have to foot the bill at the end of a bias game that he has refereed.

Going against the logic that dominates the industrial sector, our studies show that carrying out multiple maintenance tasks at once on a machine can damage its intrinsic and/or operational reliability. Once disassembled, and even after it has been reassembled, a machine will never work the same. As far as reliability is concerned, regrouping tasks is irrelevant, even with the argument that "production would be often interrupted otherwise." Our studies objectively show that scheduling maintenance while accounting for risks leads to better and greater production, with a higher MTBF (Mean Time Between Failure) and real expenses keeping as close as possible to plan—contrary to what intuitive logic might suggest.

The following example illustrates this point well (Figure 6.3). In a mining company, we chose to modulate the preventive maintenance periodicity of a truck off the road: a giant, worth at least three million dollars, with a load capacity of 120 tons. The maintenance tasks were carried out every 20 weeks, at a total cost of $8000 ($400/week). We chose to shift maintenance to every 30 weeks.

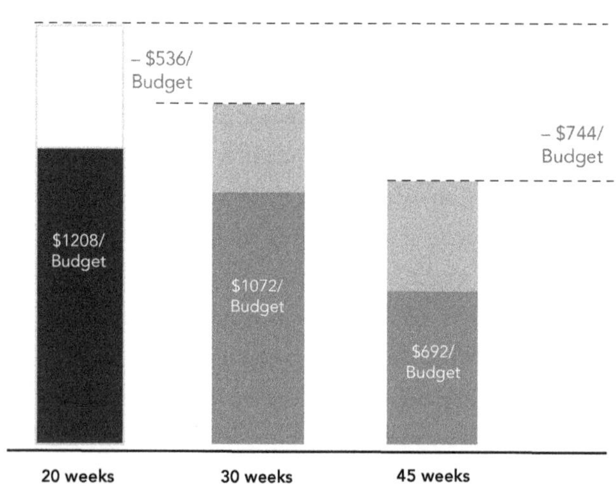

FIGURE 6.3 "Leap one..." versus optimizing the maintenance periodicity

In this specific case, we reduced the costs as well as the risks. Why? Because completing maintenance every 20 weeks was unnecessary and damaged the reliability of the engine. Since the goal was to cut the costs to a minimum, we kept running the simulation and increased the interval to 45 weeks. The costs (or, rather, the impact on the annual budget for these vehicles) were cut by almost half and the risks remained at their usual level, which the company considered to be satisfactory.

This example shows that there is some validity in the argument against dissembling too often or fully servicing a machine. For some tasks, reducing the frequency can mean reducing costs and risks simultaneously. But you don't get there by simply "postponing."

My intention is not to criticize the attitude of the maintenance manager, because I know full well that he might have the best intentions; the problem is how he implements them. While the intention is correct, the form it takes isn't. "Postponing" induces a "maintenance while we're at it" practice, based on the "try and fail" approach, which is all too common in our profession. This impedes the sound flow of information between production (who rightly wants to keep machines running), management, (who rightly needs to aim for profitability), and maintenance, (who rightly looks to ensure the sustainability of the machine). If the three don't get along, it is because they do not systematically share the data and information that would allow them to objectively analyze the impact of their decisions, the possible alternatives, and the most favorable scenarios. Instead, they give the impression of arguing on an "I think so" basis that is exceedingly random. As long as they stick to it, machines will keep on breaking down at an increasing rate and maintenance costs will keep running over budget.

It is necessary to consider the logic of production differently. Reliability, quality and production volume don't come about by decree, by decision, or through the sheer will of general management. The logic of production must be organized, built up, and calculated. Asking industrial companies to function on decrees is useless because the machine will ignore them. If we don't work the machine properly, it will send the pulley on the roof of the workshop and the production line stops.

Thankfully, there is a way to treat the machine well, with reduced costs and risks and a better MTBF. There is a way to redistribute maintenance tasks while making real savings: that is the simultaneous analysis of costs and risks. This analysis owes more to evolving decision criteria and behavioral changes in industry than to the latest technologies in intervention. In fact, it requires no upgrade in tools. Any maintenance manager is able to supply the objective arguments required to obtain a real cut in costs. He will only need to pen his data down on a sheet of paper and bring it along to the next meeting with the head of production and the general manager. This professionalizes the practices of industrial Asset Management and supplies general management with key decision-making elements, and they will then be tasked with arbitrating between stopping a machine for maintenance or keeping it running. Their decisions, whatever they may be, will be taken dispassionately, according to professional and objective criteria, and with full knowledge of the potential outcomes.

Combining these various problematics, how can one shine a light on the alleged controversy opposing the roles of Asset Management and maintenance?

One must first understand that a sum of allocated objectives does not suffice to realize a better extraction of value from assets over their life cycles. On the contrary, through their "parting" or segmenting effect, these objectives may turn out to be counterproductive as they incur the risk of creating undesirable competition between different segments of the organization that would benefit from collaborating. Indeed, as we know, if an organization is structured in a non-aligned fashion, or "in silos," then each silo may prosper and realize its own objectives to the detriment of every other segment, and often do so in good faith. However, we also know that the extraction of value depends on a global, and not segmented, success. No doubt that many among our readers will recognize, in this matter, a flawed characteristic of the organizations that they have known or rubbed shoulders with themselves.

However, other points of tension persist in this alleged conceptual conflict. This tension derives from historical and technical rationales, and one may, in good faith, recognize a degree of ambiguity in the matter. It is therefore crucial to reflect on these points of tension and to pinpoint their implications in order to overcome them.

- Let us consider, for example, the nature of the relationship that ties operational agents to the equipment (or assets): maintenance operators and Asset Managers have a direct impact on the equipment, in considerable proportions. One could even go so far as to assert that along with engineering, maintenance is the most prevalent source of impact on the asset.

- It would also be ill-advised to neglect the common career path that Asset Managers nowadays tend to share. Indeed, our discipline was primarily developed and promoted by a number of professionals originating directly from the ranks of industrial maintenance, and these "defectors" still make up a majority of contemporary Asset Managers. As a matter of fact, when I was busy, 15 years ago, presenting my first Asset Management lectures in front of more skeptical fellow engineers, I was forced to introduce these lectures as *maintenance*-orientated in order not to upset susceptible sensibilities. That fact shows the tenuous and permeable character of the conceptual border between the practices of both disciplines. Nowadays, it can be observed that agents realize the functions of Asset Managers within their organizations while they are still assigned the role of maintenance managers. This is not intrinsically harmful, but over time these ambiguities must give way to a true organizational acknowledgment of the need for planned strategic alignment.
- Furthermore, it would be irrational to neglect the considerable input from operational agents in the overall evolution of our sector. Through their advanced technical capacities, engineers have facilitated the birth and development of Asset Management. In a quasi-Freudian sense, one could assert that Asset Management seeks to add a "*business engineering*" credential to professionals who were previously known for their deployment of technical expertise. In short, Asset Management offers engineers the recognition that they previously often lacked as competent "managers." However, one should not deduce from this last point that the natural evolution from maintenance engineering to Asset Management is equal to a Freudian betrayal.
- Finally, one should always keep in mind that the intense complexification of industrial installations has come with vast consequences. Over the last decades, technological and material innovations combined with the revolution of information systems have endowed industrial installa-

tions (and particularly capital-intensive installations) with an unprecedented level of complexity. The "technical engineers" have therefore had to keep up with the acquisition of a number of diverse and often interlinked skills and capacities.

Undoubtedly, the maintenance sector had begun adapting its practices to modern industry prior to the emergence of Asset Management. Indeed, the successive transformations taking place in installations are responsible for launching a much-needed process of updating the typical maintenance practices. One can even observe that the capacity of this sector to evolve and adapt is an indicator of its contemporary proximity with Asset Management, a naturally systemic discipline that allows operators to obtain a holistic vision over the functioning of complexified installations. As we are now well aware, it is this global vision that makes the extraction of value possible since it allows the agents to develop coordination and alignment strategies for each of the different segments of the life cycle.

Through this demonstration, we have tried to bring to light the importance of each of these two disciplines, and to insist on the fact that Asset Management is neither the modern name, nor the modern nature of industrial maintenance. And yet, we assert that it would be regrettable to turn a blind eye to the common values shared by these disciplines. As we're well aware, the extraction of value is made possible by reaching a balance between risks, costs, and opportunities. Maintenance is an important factor of contribution in the implementation of this often-paradoxical equilibrium. However, as we've demonstrated, this balance is insufficient if it is established in the strict frame of an organizational segment like maintenance. The notion of setting our gaze on the entirety of installations' life cycles is therefore crucial to our activity. This specific gaze allows us to realize the optimal usage from equipment and constitutes a logical and necessary input to the action that has been practiced and designed by maintenance throughout its history.

CHAPTER 7

Operational Reliability

Before going into more detail on the subject of Asset Management's innovative conceptions regarding the economic translation of operational performance, we must discuss a crucial matter: that of operational reliability.

Reliability studies in machines what biology studies in living things. It is common knowledge that during their first years as medical students, aspiring physicians dissect corpses in order to perfect their knowledge of anatomy (the "material" aspect of the body) and progressively become versed in physiology (namely the set of functions that allow the interactions of the body's various organs). These two disciplines allow us to access a more thorough understanding of biology, which is the study of the living. Allow us to establish a parallel between this approach and the one we promote, concerned with the reliability of equipment, since those that wish to understand the *biology of machines* must also come to terms with its functional implications. It's one thing to consider assembled parts or equipment made from parts and subparts (in other words, to study assets); it's another thing altogether to understand the functionalities that drive them.

It is in this nuance that we place the importance of the following reflections.

Indeed, one must grasp that unlike students of medicine, neither engineers nor technicians in the operational and maintenance worlds have benefited from "physiological" teaching in the course of their studies. Theyare generally brilliant technicians, well acquainted with each of the mechanisms that make up the various parts of the assets, but they mostly come from engineering schools that as a whole have been slow to incorporate in-depth functional analyses within their curriculums in order to take into account the inputs of works on asset physiology. Hence, the industrial world witnessed in the aftermath of World War II a multiplication of "reliability" approaches mostly informed by behavioral models or function/dysfunction laws based on statistics-driven approaches. This proliferation was not a fault in itself; in fact, it answered a very severe need expressed by military leaders in response to the poor performances observed during the war. The industrial world has known, in recent decades, an extraordinary surge in the usage of reliability software of operational application. A considerable proportion of agents have therefore taken it upon themselves to study reliability through the means of statistic approaches, relying on increasingly powerful and elaborate tools, all of which have indeed encouraged a real progress in our understanding of the "biological" behaviors of industrial assets, but only at a severe cost.

In order to develop this notion and to proceed to a demystification of the usage of reliability data within industrial practices, I will take the liberty of relating a misadventure that I witnessed a few years back.

While on a professional trip in the Middle East, I was invited by the general manager of an industrial operation to take part in the monthly managers' meeting of the plant, which was set to take place later that day. I wholeheartedly agreed, and soon found myself sitting among the executive committee while the department managers took turns presenting their monthly balance sheets. Soon enough, it was time for the maintenance manager to defend his own results. He began by projecting onto the digital board a mathematical curve whose shape I recognized but which, I assumed, meant very little to the other members of the committee. The display was of a functional curve known as a "Weibull" curve. Hence, it represented a few dozens of dots, aligned on both sides of a line that any reliability expert would have recognized in an instant. Before he even had a chance to complete his presentation, the general manager interrupted him and asked:

"Sir, these black dots spread around the curve that you're showing us, what do they represent, in the end?"

Keeping his cool, the manager replied calmly:

"These dots? They represent the failures, boss."

At this stage in the conversation, the general manager had a reaction which I first believed I'd misinterpreted because of his somewhat clumsy English. Once he'd repeated his question, however, I could no longer doubt. What he was asking the technician for was the nature of the job inscribed on his paycheck. The unfortunate man had no choice but to confirm his station to his superior, who promptly pressed on:

"And in your opinion, what purpose does a maintenance manager serve in this kind of industrial installation?"

Trying hard not to lose face, the maintenance director sputtered something about how *in fine* his task was to avoid or to mitigate the failures impacting the site's equipment.

The general manager's reply was a scathing one: "In this case, I'm very concerned about your performance, since according to the number of dots on this screen it would seem that you have forgotten that you're here to prevent those failures that you appear to spend your time counting." A dead silence followed, after which he resumed: "As a matter of fact, I'm truly quite worried about you, as if you've judged it acceptable to index dozens—maybe hundreds!—of failures before coming here to try and convince us that your maintenance is even remotely efficient, I'm afraid your performance is simply not up to the requirements at hand."

Witnessing this scene proved to be a milestone in my career path. Indeed, regardless of the level of expertise mastered by the general manager on the concepts and applications of reliability, the questions he'd asked had brought into light a truly crucial element: real-life performances of plants and assets are rarely translatable into mathematic models. Furthermore, a field professional, however competent in his statistic approaches, will be powerless if he loses touch with the functional reality of his equipment and of the common observational sense that must dictate his diagnosis and his strategic response. Hence, the mistake made by the maintenance manager lay not in his methodical surveying of the situation, but rather in his choice to list all of these failures rather than to modify his maintenance plan.

This case, which may seem an isolated one, is in fact very widespread. In my professional life, I'm brought to visit at least 20 of these industrial installations every year across the world and very often find myself confronted with similar pitfalls. One could therefore assert that despite all of the conceptual inputs of reliability science, it has entered the industrial world "through the wrong door." Hence, industrial managers, likely encouraged in this direction by the software market, have endowed their departments with various reliability-assessing tools but have thus neglected the field of understanding the physical behaviors of industrial equipment over time and the tangible means through which one can manage the performance of assets on a daily basis via reliability concepts and models. Even in the context of the widespread usage of statistic models (which I particularly appreciate), it seems clear that a great number of managers have become entrapped by the quest for graphic or figured results and have therefore found themselves validating conclusions that defied common logic.

In my previous book, I had already hinted at this thematic, but I'd come across it through another line of thought—that of the historical habit of industrial operators to depend on failure histories to develop their maintenance strategies. This notion will become quite clear with the help of the "rearview mirror and windshield" metaphor.

7.1 Reliability: "The Rearview Mirror and the Windshield"

To summarize this parabola in a synthetic fashion, this metaphor reveals how operational managers have become accustomed to using failure histories as a systematic manner of managing their assets and coming to the right decisions. Yet, over the course of recent years, two major decision-support trends have appeared in separate professional fields and are beginning to converge in an increasingly clear crystallization.

One of these two trends is based on the concept of "uncertainty management" in decision-making, a vastly discussed notion within business schools in the context of discussions on industrial leadership. The other relies on a principle of *evidence-based* management derived from the evidence-based medicine school of thought, which has been developed by the medical body in the last 20 years in order to improve the acuity of medical diagnoses by relying directly on information harnessed from the living world. It would be superfluous to demonstrate that unlike (fatally silent) machines, the diseased are able to put their ills into words and therefore to facilitate the physician's diagnosis.

Thanks to these two approaches, we can in good faith operate an analogy stating that optimal Asset Management decision-making should take more notice of scenarios defined by looking *ahead* (or, in other words, *through the windshield* and the strategic vision it provides) rather than focus on *the mirror*, which provides technical data from the past but whose pertinence will necessarily be limited in terms of elaborating a strategy for the future (see Figure 7.1). One could push this line of thought to its absurd conclusion by demonstrating how a car mirror always displays the highway one is driving on as serene and smooth while a windshield (even if it were it tainted by dust and grime, which embody so many uncertainties that the comforting perspective of a mirror may never display) will allow the driver to anticipate or avoid the heavy truck or the tricky bump looming ahead.

FIGURE 7.1 Browsing the history through the "rearview mirror" or looking ahead through the "windshield"

It would obviously be erroneous to assert that mirrors (or, more prosaically, technical facts related in a history of failures) are not an important element of decision-support. Indeed, to go along with our previous analogy, if someone were to invite me to drive a mirror-deprived car I would categorically refuse. A windshield is needed if one desires to see the heavy-duty truck a few hundred yards ahead; mirrors are equally needed if one wishes to overtake the truck safely.

Histories of failures have always captivated operators and maintenance managers due to the latter's firm belief in the fact that understanding past elements would allow them to be more knowledgeable about future occurrences. This is true for the basic function of an industrial installation. However, as we've previously pointed out, behaviors and performances rarely reoccur identically over the course of an asset's life cycle. This, in itself, is nonetheless a very minor contradiction to the amazing credit and practical faith that operators place in the usage of histories of failures. If this was all that we could find to blame in the operational fascination with technical histories, then the issue would not be worthy of mention.

However, the study of reliability has allowed us to translate operational risks into physical and mathematical terms, which constitutes a major step forward in our understanding of behaviors and in the economic management perspective derived from this understanding—but we'll get back to this topic further down the road.

Thanks to reliability studies, the industrial world has become aware of the fact that the nature of operational risks was evolutive over the assets' life cycles, a discovery which, in turn, led to four important conclusions:

- Distinct strategic decisions can and must be elaborated in accordance with the situations of the assets themselves within their life cycles. Even if this assertion may seem self-evident, one must keep in mind that until recent years a machine was perceived as a machine, a part as a part, and the manners in which these items would have operated and been treated would have remained identical regardless of their age and maturity.
- The stakes of reliability improvement have gained in room for maneuver, thanks to the shrewdness of analyses made possible beyond the variety of elements provided by the life cycle.
- The industrial world has been able to clarify the distinction it makes between the physical and mature lives of the assets. Obviously, reliability studies, followed by LCC, which appeared in the late 1970s, played a major role in this trend.

- This has allowed the "levelling" effect of the instantaneous failure rate (lambda) induced by the effective realization of a preventive maintenance to be demonstrated beyond doubt.

These four points strike me as central in the context of proceeding to a demystification of the usage of the notion of reliability in the industrial world today, and even more so in terms of understanding the link between the behavior of assets and the treatment of past information or future scenarios.

Let us resort to using a classic example: the illustration of the use of histories of failures in the "methods" department of industries across the world.

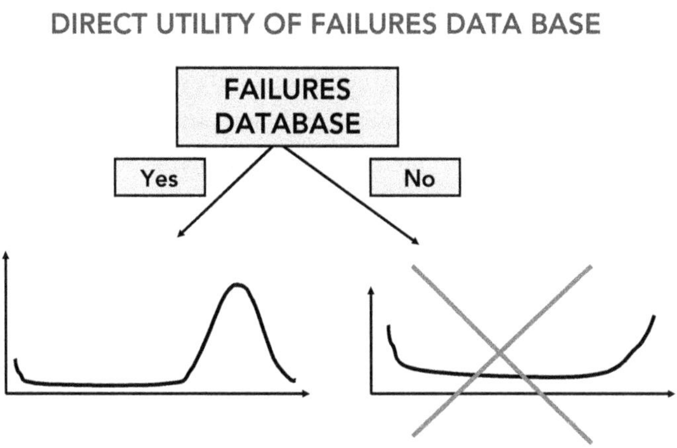

FIGURE 7.2 The pitfalls of resorting to the failure history

Figure 7.2 relates how otherwise competent engineers trace curves that represent the level of failure probability of specific families of equipment thinking that they have established the famous "bathtub curve." This is a frequent and oft-observed pitfall within organizational practice, as an unfinished probability density curve shares the overall shape of the "bathtub." However, we know that the probability density curve translates the exact distribution of the failures observed in a family of assets throughout its life cycle, while the bathtub curve, or instantaneous-failure-rate curve, cannot be deduced directly from failure histories as it is a curve of conditional probability, issued from the partial derivative of the probability density curve (obtained by observing the failure history) by the reliability curve of these same failures (which may not be obtained by observing the history).

This frequent confusion between bathtub curves and failure probability density curves would not be so consequential were it not so recurrent and so hard to observe, from a managerial perspective based solely on the organizational reports.

Even though some managers pay more attention to the derivative of the produced indicators (or, in other words, to the trends of a serial) than to the indicators themselves, the fact remains that as soon as these values (whatever they may be) correspond to a feared event with serious stakes and consequences, managers will need to be able to discuss precisely and acutely their physical (and/or economic) interpretation of these indicators in absolute terms; there lies, in fact, the only realistic path towards the construction of a sustainable solution.

A very simple, and yet often-neglected observation bearing on reliability practices brings us to another interesting point of demystification. In our sector, one must always keep in mind that assets exist in the operational world with "paired" performances. If I deem it necessary to mention this in this chapter, it is only because this very simple notion is of precious assistance on the practical side, not only for operators but also for Asset Managers. Indeed, having failure-probability density curves very close to a symmetrical format and generally focused on repairable assets implies that any point on these curves unconditionally corresponds to another more or less reliable point. For if a point is in the first median of the curve, it comes with a corresponding point of equal failure probability in the curve's second median (and vice-versa). It is therefore very useful to keep this concept in mind as for two points that share the same failure probability, since they are not in the same life cycle segment, and consequently their residual reliability may not be identical; they cannot be at the same spot on the bathtub curve. The following figures clearly bring to light how the same probabilities of occurrence can be observed at two distinct moments of an asset's group history of failures.

In an analogical process, we could compare this situation with a family in which two persons (one young, the other older) were exposed simultaneously to the flu virus ([f(t) curve]). We know that the reliability curve corresponding to this family of individuals will follow the format of the f(t) curve presented in Figure 7.3. We can easily observe that the number of survivors corresponding to the respective ages of the two individuals is not identical on this curve, while the probability of catching the flu is strictly the same. We can therefore observe directly on the instantaneous failure rate curve (lambda curve known as the "bathtub curve") that the equivalent points of risk exposure are considerably higher for the older individual than for the younger, while the cause of the disease (which corresponds to the failure of an asset) is, once again, strictly identical.

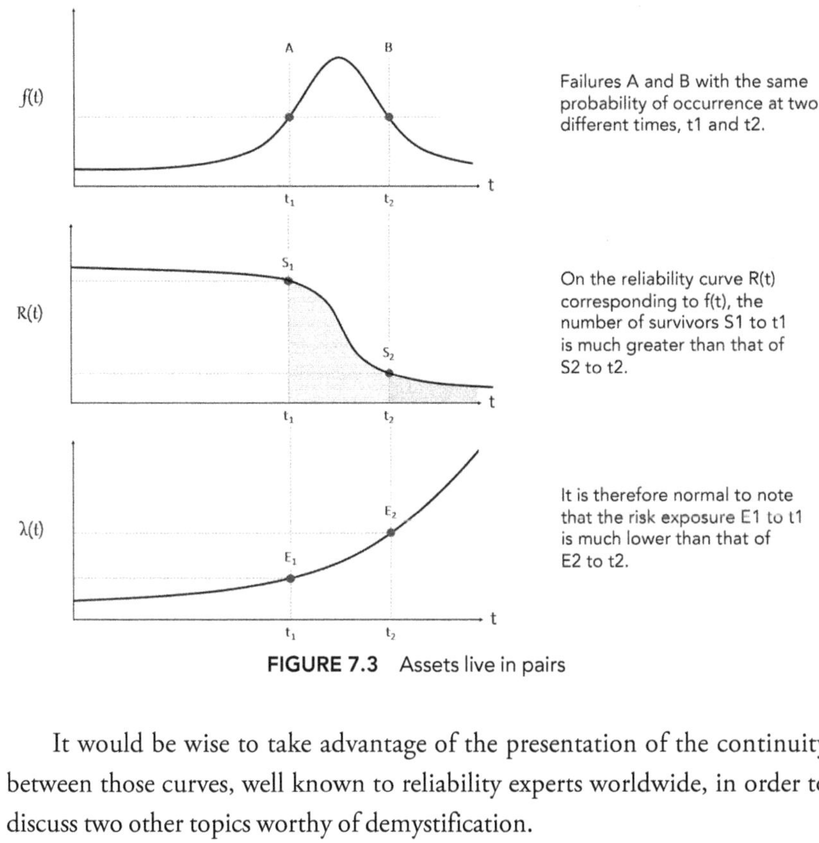

FIGURE 7.3 Assets live in pairs

It would be wise to take advantage of the presentation of the continuity between those curves, well known to reliability experts worldwide, in order to discuss two other topics worthy of demystification.

7.2 Demystifying MTBF—Mean Time Between Failures

The first of these topics comes in the form of the extra care that should be applied when dealing with the notion of MTBF (mean time between failures), as observers have noticed that frequently industrial sites produce copious amounts of reports and studies based upon MTBF values, which have been obtained by analyzing a small number of assets from an apparent same "family" or even from different families altogether. The problem at hand here is that of representativeness and statistic signification, for as MTBF is a mean, it may only have a calculative meaning (and consequently, a physical interpretation) when it is deduced from a minimal number of observable elements on the random variable (failures, in this context). Although most professionals are well aware of this pitfall, they tend to defend their attitude by stating that on this particular site they only possess a small number of equipment from a singular family, which, however true, does not excuse the imprecision of their calculations. These erroneous calculations often induce the implementation of actions that go against what an adequate reading of the reliability performance translated by the MTBF would have suggested.

Let us remember that a MTBF must in theory represent half of the equipment, and is not only the mean for equipment present on site but for the entire family of this equipment, even beyond the borders of the site. In other words, the MTBF of a model-X pump from a "Y" manufacturer must be identical if we only have a small number of X-pumps or an infinite number of these. It is an intrinsic value, and as such it demands a certain amount of rigorousness as to the minimal number of observable equipment in order to respect the required statistic signification and representativeness. I'm well aware that raising this debate will worry those among our readers who have become accustomed to proceeding in such a manner in order to calculate the MTBF of their onsite equipment. Therefore, I shall attempt to provide them with information on the ways in which to proceed more adequately in order to avoid such pitfalls. My reasons here are not only dictated by the quest for statistic rigor, but also out of concern for the countless managers who month after month open their operational reports with MTBF values that are not only inexact, but also unable to properly profit the decision makers who will read them.

What I mean to assert is that even though one can rejoice in the idea that a great number of potent software have appeared in the industrial to provide accessible means of practicing operational reliability, it would be reasonable and useful to also go back to practicing very easy graphic calculations of MTBF in order to interpret more efficiently the nuances that come with the notion of "mean," which in the end is relatively weak in terms of providing pertinent elements of decision. The histogram presented in Figure 7.3 and displaying the MTBFs of a family of equipment reveals not only how easy it is to proceed to a visual assessment of the MTBF but also how it is possible to interpret issues specific to each segment of the MTBF by analyzing the gaps that exist in comparison to the adjusted curves.

The following graphs (Figure 7.4), representing the MTBF for a family of equipment, show that it is easy to calculate the MTBF visually. This improvement, both simple and elegant, was put forward by Igor Bazovsky during the 1960s.

In reliability, the MTBF in the case of an Exponential Law, is written:

$$m = 1/\lambda$$

m being the mean of the exponential curve.

It is a parameter of Time that describes the reliability of an asset or system obeying the exponential law.

The reliability function R(t), also called Probability of Survival, is written:

$$R(t) = e^{-t/m}$$

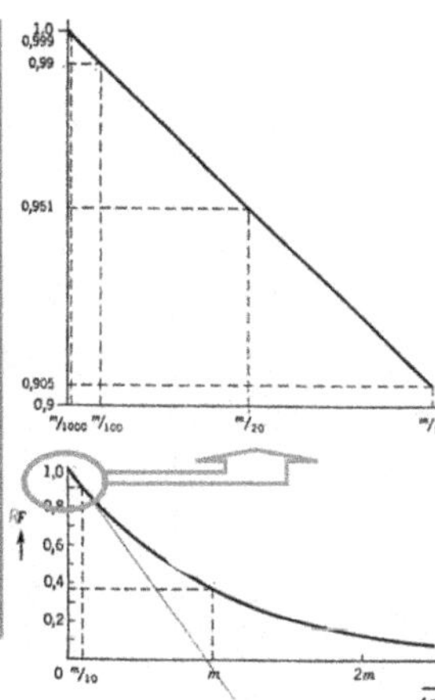

FIGURE 7.4 The exponential law and the abacus of the MTBF (Source: I. Bazovsky)

In my opinion, it is urgent that this idea, according to which one need only resort to a visual study of the graphics to assess MTBF values on a daily basis, be shared by the majority of Asset Managers and operational managers worldwide. Indeed, in the context of reliability studies, we often use, for practical reasons, the opposite value of that of the lambda rate in order to deduce the MTBF. This hypothesis is based upon the fact that this time parameter (like the lambda) describes the reliability of an asset in a system whose failure probability density would follow the exponential law. From these hypotheses, one can then benefit from the simplification, which would consist of focusing our gaze towards the highest reliability values (usually the most sought-after) and hence work with a form of abacus that would make the calculation of MTBF for the most reliable pieces of equipment a linear calculation. Even if this process is less sophisticated than the many types of software that enable one to calculate MTBF, it is a practical and empirically efficient method. Furthermore, I have an intimate conviction that the calculations obtained in such a fashion are not only more precise but also more easily operable, while they render the physical interpretation of the operational reality much more on point.

Once more, if one wishes to strive towards the extraction of value in Asset Management, one should rejoice in every approach that allows us to reach information relevant to decision-making in the simplest and most accessible ways.

7.3 Reliability and Professional Behavior

If machines could talk, they would complain about the maintenance technician over manipulating them. For various reasons, detailed further, the technician cannot stand to see his beloved machine deteriorating, even less so having failures. I call this attitude the loving care. It is very widely spread among maintenance technicians and occurs frequently in the daily life of factories.

In the mind of the technician, who is paid to keep the machines operational, the loving care must seem like a touching demonstration of zeal and professional aptitude. In fact, this is how it looks—but it is misguided. The loving care leads precisely to what we would want to avoid: deterioration, breakdowns, and losses. Consider what a machine would say if it could talk:

> *"Please, don't do it, you are fiddling with me again! Is this the way it is going to be now, every week you open me up, you take me apart, you hollow me out? Don't you understand that it is painful? I quake on my axle every time I see you coming up to me with that toolbox.*
>
> *"I used to think you were my doctor, and I liked it when you took care of me. Now, you are more of a monster or a sadist, the way you turn me upside down. You went too far last time when you rewound my engine. If I were old, fair enough. But I am still young, and my engine is rewound enough already!*

"I'd like to see you handle it. I wonder how you would react if your doctor were to carve you open every week, just to make sure you don't have an ulcer. This is exactly what you do to me: you carve me open to avoid deterioration and failures. You don't even wait for the symptoms to show. If you think that is useful, you are fooling yourself.

"Have you forgotten that I was designed to produce 10 tons a day? Now, just because the Chinese are desperate to purchase what I make, I must produce 13 tons. You should be defending me, but instead you pretend that it is without consequence. I am damaging my health, trying to keep with your bosses' pace, so there is no point for you to keep replacing my parts. What is the point of replacing the cells of my sensors when they start blinking differently, of replacing my bearings or of rewinding my engine (how humiliating!) ahead of time? It is putting me in a state! My health is worsening, and, at that rate, I will last half as long as I should.

"If it keeps going like this, I will have failures even if you take care of me (at least that is what you think you are doing). I will break down, you will fix me, I will break down again, you will fix me once more ... until someone decides that I am useless and throws me in the trash. I've seen it happening over and over again in this plant!

"You should know that this is a death sentence. Have you no remorse?"

To understand how a seemingly positive attitude can lead to disaster, let us start this chapter by introducing the concept of intrinsic reliability, from which both loving care and maintenance originate. To keep things simple, let us imagine that intrinsic reliability is the DNA, or genetic code, of a machine, intertwining two key characteristics: capacity and performance.

As you know, a machine exits the production chain with specific characteristics of capacity (produce 10 tons a day of a given product, for example) and performance (maintain this daily production rate over ten years, let us say). We set it up in the plant and turn it on. We assign it a preventive maintenance program, and all is well for a time.

But, because of fierce competition or a surge in demand, we might decide to increase its production capacity. This will necessarily affect its performance and its output. If a machine that was designed to produce 10 tons a day suddenly increases to 13 tons, the performance of its life cycle will tend to drop below its original 10 years. Imagine that you have a felt tip pen designed to draw a line that is 0.5mm wide (capacity) over a kilometer (performance). You decide that you want the line to be 0.7mm wide. As you know, you only have to push harder on the pen to widen the line. But, under such circumstances, will the pen be able to sustain the line over a kilometer? That is the question.

The relationship between capacity and performance is easier to grasp when compared to communicating vessels (Figure 7.5).

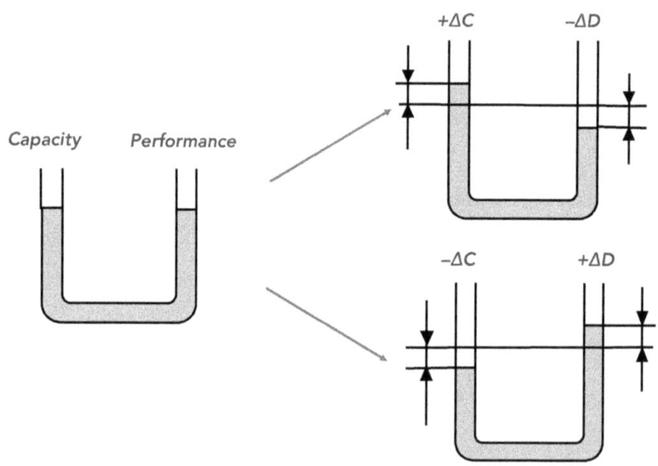

FIGURE 7.5 Analogy between reliability and communicating vessels

In an intrinsic reliability situation, the capacity and performance levels are perfectly on par. However, when the capacity level increases, the performance level drops proportionately (and the other way around).

In other words, increasing performance beyond the DNA-set limits of the machine will inevitably affect its performance. Yet, the director seems to conveniently suffer from memory loss as he summons the maintenance technician: "I was asked to enhance the production of this machine—its capacity must increase by 30%." Upon hearing these words, which ring like a direct order, the maintenance manager will try to touch upon the DNA of the machine (its intrinsic reliability characteristics) and reply: "With all due respect, sir, the machine is already running at full capacity, and even above it." With a pat on the back, the director will only add: "There's another reason to take good care of it!"

And this is where the loving care comes in. While the maintenance manager is perfectly aware that increasing the capacity of the machine will trigger an imbalance, he does it anyway. He is in no position to dispute the "orders from on high"—although, as far as I am concerned, outlining their consequences to management is a "professional obligation." Since the smooth running of the machine is his responsibility, he duly looks after it. The affection he has for the machine combined with the huge pressure imposed by management will lead him to double down on the care he shows to his protégé. He will increase the frequency of preventive maintenance, replace parts ahead of time, never leave the wretched machine alone, etc. He will take the opportunity of every breakdown to carry out a complete servicing of the machine. This is another widespread reflex among maintenance professionals: "maintenance while we're at it." "While we are checking the belts and pulleys, I might as well check the bearings. While we're at it, I might as well service the mechanical part, and the electrical part," and so on. Thinking back on these practices, which were my own in the past, I wonder if I was looking after the machines or after my own job!

For a while, the machine will put up with these circumstances and keep running. But, as the increase in capacity triggers a slump, it deteriorates increasingly fast. It begins to have failures earlier than expected—even the most able and passionate loving care can do nothing about that.

Any machine undergoes deterioration, even when running at the intrinsic reliability level (Figure 7.6). It might be a slow process for some (see curve A in Figure 7.6); it will be brutal for some others, especially those with a weak DNA (curve B); for others still, the deterioration will be subtle (curve C). Like any man, a machine starts dying on the day of its birth. This truism is captured by a coinage from our German colleagues: they prefer the expression "useful reserve" to "intrinsic reliability." Any running machine drains its useful reserve, until it becomes useless.

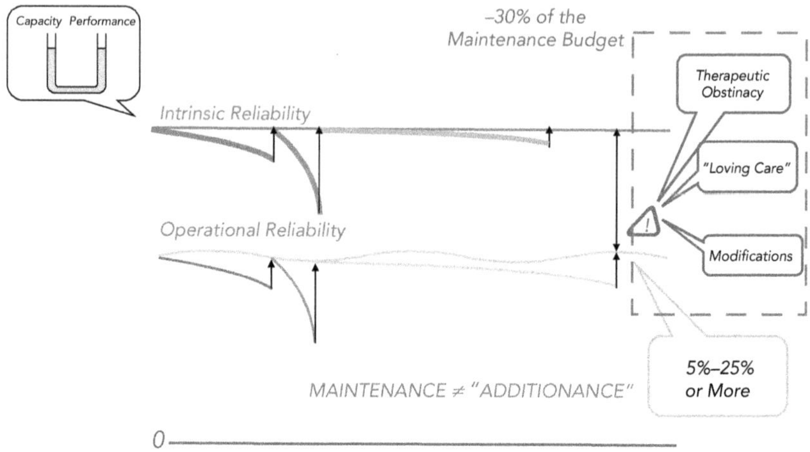

FIGURE 7.6 Distinguishing between intrinsic reliability and operational reliability

Every machine on the planet meets the same fate: the junkyard. The length of their deteriorations is the only variable. We must never forget that the maintenance manager is always on call to ensure their optimal performance. His job is to impede the rising curve of deterioration and to keep the machine on intrinsic reliability levels, in terms of both capacity and performance. In Figure 7.6, maintenance operations are shown by vertical arrows: they indicate the "rejuvenation" triggered by preventive maintenance against the natural wear of any industrial asset.

However, not every damage is elastic or repairable. Some are plastic and permanent, despite maintenance's best preventive efforts. Here is a further similarity between man and machine. Notwithstanding the good care that my mother had for me, I was short-sighted by the time I was nineteen. My eyes deteriorated at the "plastic" level: I could be using the best eye drops in the world, but my eyesight would still be impaired. The same thing happens to 5–15% of machines in any plant: one day, they become "short-sighted." Their "plastic" wear brings reliability down to a level of "operational reliability," shown by the serpentine line in Figure 7.6.

Once again, loving care rears its ugly head. The maintenance manager refuses to accept the change in behavior of his dear machine. He would do anything to bring it back to normal, that is, to its intrinsic reliability level. He opens the machine up, replaces a part, and rummages around. He might not know that a specific part of the machine has deteriorated. He might deduce it, but he will not know which part. Or, he might be perfectly aware of it but unwilling to accept it. What will he tell his boss, who trusted him with this machine? If he recognizes the "plastic" damages, wouldn't he also have to recognize his own incompetence? It's a firm no—he will not bear the new condition of his protégé. He'll feed it some eye drops!

Now, are you clear on what the loving care implies? It is a form of industrial Asset Management that disguises, under apparent skill and professionalism, a weak spot through which profits and benefits leak.

The maintenance manager, whose role is to ensure the smooth running of the machine under any condition, is at the heart of it. He follows orders regardless of their silliness—no, he bends to orders. It never occurs to him to say: "please be aware, as director, that you are bound to choose between increasing the machine's capacity to 13 tons and having it last 10 years." Besides, there would be no point in disputing an order on a feeling or an "I believe that …" basis: it requires demonstration. Winning over the plant director takes hard data, objective arguments based on facts, not rumors. Furthermore, even if he has enough to build a case, the maintenance manager might not know how to organize the data.

The maintenance manager therefore has no choice: the machine must run under such conditions, and it must last; that is his duty. Being the competent professional that he is, what does he do? Absolutely everything. He never leaves it alone, much like a doctor who carves his patient open every week under the biased belief that an ulcer will inevitably show sooner or later. He loves his dear machine, and he takes pleasure in disassembling it, replacing parts, and reassembling it, in the name of its health.

During my time as a consultant, I have encountered unbelievable examples of loving care. The one that I am about to report deserves to be collected in dedicated anthologies, and it perfectly illustrates how far this bond between man and machine can go. In an energy production unit, there was a coal-fired power plant. An equipment commonly known as a "scraper" ran under the furnace; it resembles a treadmill, designed to evacuate the ash from the furnace's internal combustion. Scrapers are mounted with tens of metallic plates ("brushes") that collect the ash from the bottom of the furnace and convoy them to repository. Could anything serious happen when one of the brushes deteriorated? Not really, since any dysfunctional brush was covered by the following one. There were, I insist, tens of brushes dedicated only to collecting the ash.

When asked to optimize the maintenance of this plant, I naturally concluded that there was absolutely no need to perform preventive maintenance on the treadmill. The equipment was stopped, disassembled, the brushes cleaned, and the damaged ones replaced, much more often than was necessary. Why, since the wear of the treadmill had no chance of impeding the smooth running of the plant? Furthermore, what mattered to the company was not to stock ashes but to produce kilowatts. You have here a typical case of spending thousands of dollars and getting nothing out, not a single cent. Therefore, my project suggested to simply delete preventive maintenance on the treadmill.

My advice was followed and implemented. When I visited the plant a few weeks later, I was summoned by the industrial director, with whom I would usually have little contact until the project was completed. I wondered, "What have I possibly done wrong?" But, as I discovered when I reached his office, the director was all smiles. He poured me a coffee and put me at ease. After a while, he shut the door and announced that he had a story to tell. "Listen, I am a fan of your work, I have no complaint on that front. But scrapping the preventive maintenance on the treadmill puts me in a tough spot."

The director then explained that the technician formerly in charge of maintenance on the treadmill was sneaking into the plant every Saturday afternoon to take care of it. "This is a security issue, we have to ban him from the premises," he went on. "On the other hand, I wish to avoid conflict. He is an exemplary technician; always has been for the thirty years he's been working with us." To conclude, the director kindly asked me to revert to preventive maintenance on the scraper, which, given the situation, seemed fairly harmless.

Can you believe this story? The maintenance manager could not bear to let go of the scraper. With the passion of the man who rinses and polishes his car on a Saturday afternoon, he would sneak into the plant to uphold the maintenance job that had been taken from him. I try to imagine what went through his head when he first heard that his duty had been scrapped: "For years, I have been shining every brush, and working hard to keep the scraper good as new. I cannot bear the thought of it getting damaged! This equipment is like a son to me. Besides, everyone will blame me if it suffers damages—they'll say I am no good with a scraper."

Do you get it? This is what I mean by loving care.

Such passionate attitudes highlight the pride that shines in the heart of every maintenance man. The way he sees it, avoiding machine breakdowns is not only his livelihood but also a matter of honor (D'Iribane, 1989). Do you seriously expect him to let his protégé suffer breakdowns on his watch? Not in a million years. Could he accept it if his machine suffered plastic damage and was suddenly unable to produce as much as before? Of course not. Do you expect him to dodge when handed one of these impossible missions? You don't get it. The machine must run, by will or strength. If it doesn't run as it should, or if it fails, he will be held accountable. It will be his fault.

Driven blindly by his pride, the maintenance technician is prone to truly heroic gestures. He will work overtime if need be. To conceal how much work goes into the machine, he will repair the damaged parts and store them in a drawer alongside his toothbrush, thereby saving himself from the humiliation of having to request spare parts from the warehouse. He is close to saying: "I thank you, mister director, for trusting me with the delicate task of ensuring maintenance on this badly damaged machine, for, after twenty years in the field, I am the only one fit for the challenge." Thus, the maintenance technician turns into Superman, fitted with superpowers. He would do anything to keep his machine running.

Although the orders he bends himself to are absurd, he derives a sense of pride from them. He is sure to be rewarded for his brave and dedicated actions. Besides, criticizing his sense of dedication would be the worst thing to do. Would you criticize the maid if the sink was too shiny? Do you tell her she works too much? Well, the same goes for the maintenance technician.

There is one scenario that I've come across often in companies: loving care regarding machine bearings. Ever since the device that diagnoses the wear of the bearings by reading their vibrations was created, taking them apart has been superfluous. Doing a reading periodically is enough; when the vibration harmonics indicate possible damages, the time has come to replace the part before it fails.

And yet, there is always a maintenance technician to question the reliability of the device, who insists on inspecting the bearings with his own eyes. If a sign of wear detected during a first reading has slightly increased at the next reading, he will not wait for the third reading, regardless of the leeway. He runs to the warehouse, grabs new bearings off the shelf, opens the machine and replaces the parts. Why wait upon a bad sign at the next reading? What if the bearings broke before then? He will not take such risks with his machine and expose himself to blame from higher up. If anybody objects that the part was replaced early, he will have two readings to show to support his decision.

This is what I mean by loving care: the ways in which the maintenance technician shows his affection to his work tool, and the means by which he tries to avoid the deterioration of the machine being correlated to his own professional expertise. The attitude he demonstrates in the process, however, is everything but "professional expertise." To escape any suspicion of incompetence, the machine would have to be forever in a pristine state. Since deterioration is inevitable, he replaces parts on the belief that he is anticipating future problems. Keep risk to a minimum, right? Besides, replacing parts is what he does best! Between replacing bearings and drafting a report outlining the economic benefits of wearing them out a little longer with no added risk to the machine, he will consistently choose the former. This is the unbelievable paradox that came with the formidable progress of conditional maintenance techniques, especially vibration-measuring techniques.

How much longer must we suffer having maintenance technicians, from the base worker to the most experienced engineer, spend the best part of their work time on tasks as basic as systematically replacing parts? The extraordinary progress achieved in the past 30 years, in terms of training maintenance professionals, must not be summed up by an infinite supply of "luxury part replacers" flooding the job market! If maintenance managers continue to spend, as our benchmarking shows they do, most of their work time supervising workers and foremen—a job that any least-skilled employee could easily take on—they will continue to impede worthy managers from stepping back from daily maintenance operations, even though their time would be more usefully dedicated to the continuous and lasting improvement of their function, in the mid and long term.

In this specific example, the only mitigating circumstance is that a broken bearing would cause serious damage to the machine. However, some loving care scenarios simply have no valid justification. A while ago, I was working with a European metro and bus company. The maintenance manager was so deeply perfectionist that he would replace the lightbulbs in the vehicles before they had burned out. His verification process was, indeed, sophisticated. Since the lightbulb's intrinsic reliability promised a thousand-hour guarantee, he would have the lightbulbs replaced after 950 hours. How he calculated the life span of each lightbulb is still a mystery to me. This is madness! On top of the waste induced by replacing still-functional lightbulbs, add the sheer time and energy loss of monitoring this huge task. And what for? To prevent anyone from wondering, at the sight of a burnt lightbulb, why "the technician in charge hasn't replaced it?"

This is what I mean by loving care.

For the maintenance technician, this attitude is a sign of professionalism. It justifies his role in the company and prevents serious failures. Yet, the actual consequences of his behavior are quite the opposite. Every time he bends to ridiculous orders, for submissiveness or for pride, he goes against the company's long-term interests, and chooses the costliest alternative. If he could explain to superior management that their decisions will inevitably have negative outcomes, everybody would be better off. What I mean to highlight with the phrase "ridiculous orders" is, of course, that, as far as industrial and management strategy goes, any decision-making process based on incomplete or skewed data should be deemed ridiculous. Why is it skewed? Because the maintenance technicians with a grounding in the plant struggle to pass objective data on to superior management. In other words, until every technical manager in the fields of production, maintenance, purchasing, and stocks has dropped for good the "guesstimating" habits that still pervade the industry at the start of the 21st century, the decisions made up high can legitimately be deemed ridiculous.

Although the maintenance manager is the main character in this loving care scenario, his superior is not exempt from all responsibility. In fact, he might be fine with passing on orders that dictate an increase in the output of the machine. Why not? After all, the maintenance Superman never lets the machine down. The superior concludes that his decisions are informed—and sound, even, since nobody disputes them. He finds that it is easier to communicate with the maintenance technician in short and snappy orders: "Increase machine capacity. No discussion." Yet, he fails to realize that this attitude does nothing to improve the skills of production or maintenance in the company that he manages.

The manager's behavior, along with that of his maintenance technicians, makes up what I call the "double spiral of failure." Since each reinforces the other's behavior, the spiral is never-ending. Luckily for both, the excess spending, problems, and losses—wherever the spiral leads—incurred by their behavior is overshadowed by the benefits inherent to their function.

As President Abraham Lincoln once said, "you can fool some of the people all of the time, and all of the people some of the time, but you cannot fool all of the people all of the time." Thus, one wretched day, things fall apart. Be it for the high stakes, the profit margins, or the "gunshots" that sometimes occur in a plant, the maintenance technicians and their superior will eventually have to wake up and face the music. For everybody's sake, this might as well happen right now—and, in the field of maintenance, today means yesterday. They simply must aim, tangibly and quickly, towards the "virtuous spiral of success" where everything goes from good to better.

CHAPTER 8

Risk and Its Management

Studies operated by manufacturing plants and public service infrastructures alike share a common analysis: during the operational phase, the utmost challenge faced by leadership takes the form of guaranteeing a functional stability.

In the case of capital-intensive industries, this implies upholding the lowest possible level of uncertainty as to the guarantees provided to customers and employees. Overall, industries and infrastructures are required to provide a reliable and efficient service, both of which necessarily depend on ensuring the proper functioning of the installation and of the assets that compose it. The guarantee we've evoked comes in many forms; of course, it is primarily technical, but one should not overlook the importance of guarantees provided on matters of security, sanitation, or environmental preservation (the famous HSE factors). Furthermore, it is clear that these guarantees generate a considerable economic impact.

Organizations must therefore maintain a balance between different criteria: productivity on the one hand, security and functionality of their assets on the other. Considering the capital importance of this balance, it seems crucial to identify clearly the common denominator to the entire set of challenges beset by these various criteria. In the Asset Management rationale, this common factor is all but known: it is risk.

Indeed, despite frequent politically correct statements from regulators claiming that "safety and environmental protection are priceless," many officials seem to forget that these values come at a cost and that since the risk may never be suppressed wholly, there will always need to be a consented level.

Indeed, every choice comes with a cost and even deciding not to choose implies taking a *risk*, which eventually will evolve into a cost. As it was established by Cass R. Sunstein in his essay *Choosing Not to Choose*, the refusal to choose between various given options is one of the most frequent decisions made by industry leaders. Yet, not choosing inherently means exposing oneself to an ulterior cost; hence, it is a risk. The modern *homo oeconomicus* is rather inapt to intuitively master the idea of risk. However, one is unable to avoid it: either one knowingly faces a risk, or one chooses not to, all the while knowing that one is clueless as to the level of risk contained in these two options. To make it worse, even when one is informed of the level of risk exposure represented by an option, one is generally unable to translate this degree into economic value. That is where the common denominator of all progress induced by Asset Management in the industrial and infrastructural worlds can be perceived. Indeed, we have developed the tools and models that allow us to monetize and valorize risk. One can therefore define Asset Management as the discipline that enables a true comprehension of risk.

Insufficient data is the root of the problem: we tend to cut costs without accounting for the imputed risks. By risk, I do not simply refer to potential failures caused by a lack of adequate preventive maintenance, which incurs the added cost of corrective maintenance. My notion of risk also includes the production losses caused by the production downtime, and the correlated shortfall—that is, the profit forgone during the pause in production. Risk thus includes a double prejudice: the machine must be repaired, and production must be stopped.

In industry, decisions aimed at cutting maintenance costs very often lead to imbalances in the plant, which in turn cause a halt in production and losses considerably greater than the expected (and ultimately wasted) savings.

A quick computation will prove that the cost of an hour's shortfall is 250 to 900 times greater than the cost of one hour invested in preventive maintenance. This should give you an idea of the consequences of poor decision-making when it comes to trimming costs off the maintenance budget.

It goes without saying that nobody would even make such decisions if risks could be accurately calculated. But here is the real problem: how do we quantify such risks? How to elevate them on a par with costs where they should be? That is, how to convert them into hard data to prove their economic impact?

This is exactly what was going through my mind when I was starting out as an engineer. I remember every minute of the day when I offered to modify one of the machines, convinced as I was that issues on this machine would otherwise build up until production ultimately had to be stopped. My boss replied: "Fine, Celso, why don't you draft a report that clearly outlines the cost, length, and profit of the operation." I quickly reached a painful conclusion: I was unable to calculate the risks incurred by not doing the modification. I was at a loss for economic and financial arguments that proved the relevance of my suggestion. I soon realized that my own personal conviction would not suffice.

This scenario was repeated over and over again. As an engineer, I was profoundly convinced that my improvement proposals were valid, but I struggled to back them up with proof. I had to stamp hard numbers on my instincts, round them up on a board and demonstrate their soundness to the company, but I couldn't figure out how to proceed. How many times have I confusedly stammered this belief during operational meetings! But it wasn't a fair fight. On the one hand, there was a cut in costs with a figure attached; on the other hand, mine, an increase in risk exposure that I could not quantify. I would walk back to my office sulking because I was certain to be right but couldn't prove it. And sure enough, what happened a few weeks or a few months later? The risks were real and derived from the machine failing or breaking down. We would go ahead with the modifications I had suggested earlier, in bad conditions and at a higher cost, which only increased my sense of frustration.

The scenario that I have just described results from a firmly rooted paradox that runs through most companies today. While maintenance and production costs steadily drop over time in theory, they consistently increase in practice. The estimated budget shrinks but the actual budget always exceeds projections. It's a ticking time bomb! The costs incurred by machine repair always run much higher than expected. Modifications, upgrades, parts replacement ... every unplanned task that has to be completed in practice digs a huge hole in the budget. A good indicator of this is the shifting moment in time when amendments to the maintenance budgets get requested. Every year, this moment comes a few weeks earlier. Twenty years ago, the maintenance department would open negotiations with management over amendments around September or October. Today, we see amendment requests coming in as early as June or May.

According to the ISO 31000 standard, risk corresponds to "the effect of uncertainty on the realization of objectives." Risk is therefore the irreducible variable that eludes control from leadership. If it cannot be suppressed, there are, however, various ways of apprehending risk.

Historically, four distinct manners of treating risk have coexisted:

- Accepting risk after the assessment of its impact
- Reducing risk by the means of asset-impacting actions
- Transferring risk toward a third party
- Avoiding risk by putting an end to the activity that generates it. This is a distinct option altogether from actions aiming at the mitigation of risk since the latter only strive for the reduction of the probability of occurrence.

If we are so concerned with risk in the context of discussing the challenges met in the operational phase, it is because its management is truly crucial. In fact, it is the inner essence of Asset Management.

Unfortunately, even though risk constitutes a true pillar of Asset Management, its management remains dependent on the level of rigor deployed by organizations to face potential risks. In other words, one can take satisfaction in the fact that the acknowledgment of the need to correctly manage risks has made great progress; however, the fact remains that even though it is absolutely necessary, the management of risks is not compulsory. In the end, organizations are not equal in the face of risks, and even less so in the face of their management. As we've said, organizations understand that risk is "priceless," but they also realize that it will always come at a cost, the latter of which will perpetually rise.

One must truly grasp that Asset Management has introduced an original and innovative process that was widely craved until recently. Over the course of the last decades, we've brought into the public sphere and in a definitive fashion the concept of monetizing risk, or, in other words, proceeding to an economic translation of operational risks into financial and economic value. To be sure, this idea is not fresh, but never before had it been so structured, so reliable, and so adapted to operational practice. Today, we are equipped with the tools, methods, and systems allowing us to proceed to the valorization of risk in a most practical sense.

Hence, we are able to respond optimally and in real time to the challenges of the industrial world.

8.1 Monetizing Risks: Stakes and Originality

Experts agree in stating that nowadays the paralleling of decision-induced costs with the risk exposure of non-decisions (which will necessarily morph into costs) is a central lever in the effort to meet the challenges set by the infrastructural and industrial world. In other words, one must be able to anticipate in an objective and informed manner the consequences of each decision made by leadership in order to reach optimal choices or, at the very least, be able to avoid having to make a choice that would induce unassessed potential consequences. How often must we hear managers enumerate in their end-of-year meetings the poor decisions taken in previous months? In order to provide the necessary degree of clarity to decision makers, one needs a language in which to express the information at hand: this language can only be that of those who make decisions within organizations—the economic language.

Quite often in the traditional corporate culture, managers make decisions dictated by the general belief that risk must be avoided at any cost. This aversion to risk is based on a limited quantity of objective and concrete information. Quite frequently, the managers responsible end up actually exposing themselves to more risk than they would have if they had based their decisions on common sense and on what they could observe if they only tried.

Clearly, what is at hand here is a blatant lack of reliable and consistent information, which can explain why managers are hesitant to base their activities on fragmented and insufficient elements. But this situation is no longer a fatality; nowadays, Asset Management provides a common mindset as well as a rationale that is both adapted to each situation and founded on pertinent information.

By way of an example, let us consider the case of economic shortfalls. They constitute a very tangible and daily consequence and they can be very useful in the context of asset-related decision-making. One cannot overstate the economic importance of such a factor: it is commonly estimated that an hour's worth of shortfall induced, in industrial organizations, by the failure of an asset, amounts to hundreds of times the direct cost induced by an hour of preventive maintenance.

In parallel, we know that an action intended to prevent a failure (and therefore, a shortfall) may also cost up to a hundredfold the direct cost of the task itself; yet despite this high ratio, it will always be justified. This is known as the opportunity factor in Asset Management. It is a strong argument and inductor in favor of an adequate maintenance strategy inscribed in an Asset Management System (cf. ISO 550001). The opportunity factor corresponds quite simply to the cost of avoiding a feared event—it is therefore universally achievable.

However, it is impossible to quantify through a purely accounting prism. Hence, managers from traditional organizations choose not to take this phenomenon into account and end up exposing themselves to even more economic risk because they've been unable to translate the initial shortfall into pecuniary terms. They tend to "bury their heads in the sand" in the face of various operational risks. However, to conclude our example, we can observe that the Asset Manager is well prepared to quantify shortfalls by means of the integrated-risk approach. Moreover, if one is able to translate this information into economic terms, then one can demonstrate that a scenario in which integrated risk is monetized is vastly preferable to the traditional scenario in which managers act, so to say, as *blind* men and expose themselves to risks that they're unable to intuitively assess.

The notion of "risk," in the Asset Management line of thought, brings us to question the relationship between knowledge and action. Indeed, the industrial world demands of managers that they "act" (or more precisely, make decisions related to the management of assets) without "knowing" (in this case, possessing the information relevant to these decisions). Reciprocally, they possess a vast amount of information that they choose not to use, simply because they prefer to resort to seemingly more efficient operational modes. Our vision of risk management, on the other hand, asserts that a decision is "valid" only if it is based on solid information and that it is therefore impossible to "act" without possessing a knowledge that, if it isn't optimal, is at least sufficiently consistent and pertinent.

In practice, one need only become aware of the flaws of an operational method and be able to quantify these dysfunctions to undertake the first steps of a risk analysis.

Overall, our approach takes the form of a call to pragmatism.

It is quite clear that organizations are by now well equipped to pinpoint the sources of dysfunctions within their industrial installations. They are also able to state the major issues harming their installations and to rate these issues within a file known as a *ranking*. It would therefore be simple to avoid implementing ineffective managerial, operational, and maintenance policies—except for the fact that organizations fail to identify such policies because the decision-making structures are not based on the true root factors of operational risks.

Everything seems to happen as though the dominant decision-making model was based on a hope of doing better or a fear of providing the necessary effort to avoid a problem, rather than on concrete information that would make it possible to truly manage or contain risk. Everything seems to happen as though decisions were still taken on the basis of inadequate elements that only have to their credit their commonness and the feeling that they provide of illusory efficiency.

It is this *wait-and-see* philosophy that encourages managers to treat risk in an inadequate fashion—and it is also in this sense that Asset Management embodies an outstanding qualitative leap. It makes it possible to go beyond the first stage of risk management by introducing the crucial notion of risk monetization. Hence, it makes it possible for decision makers to base their projects on concrete and objective elements and to analyze them in a perspective of vulnerability. Furthermore, it is thanks to its monetization that risk can be translated into the language common to every decisional agent, namely the economic language.

8.2 What Type of Management for Risk?

In spite of the many precautionary principles implemented by nations over the years, it is clear that mankind may not exist without exposing itself to a varying degree of risk. Moreover, these very precautionary principles have only met very minimal results—if they enabled us to avoid major catastrophes, we would undoubtedly have noticed. Risk is truly inherent to human existence, and consequently to the organizational experience. Fleeing risk at all costs therefore equates to fleeing life itself.

This inherent character implies that it is essential to learn to tame risk if one aspires to progress. This is particularly true in the context of Asset Management since mastering risk makes it possible for managers to proceed to the extraction of value.

In order for organizations to realize the optimal extraction of the operational value of their assets, they need to be informed of the various scenarios at hand—in terms of risk/cost and CAPEX/OPEX balances—which will in turn allow them to develop successful strategies. Indeed, accepting the inherent character of risk does not imply that one is willing to endure its effects; on the contrary, our idea is that risk must be mastered in order to turn it into a source of profit. Hence, Asset Management has brilliantly introduced the idea of a measurable and *calculated* risk, through the means of its *economic translation*. Thus, a new discipline has emerged, which allows one to guarantee a projected level of availability at an optimal expense.

However, we must still examine and define the adequate methods and fields of application for these risk-related tasks.

Historically, the first vectors of progress brought by Asset Management to the assessment and monetization of risk were focused on the determination of optimums.

Let us consider the case of an asset that would be poorly maintained on a daily basis. If it is neglected for some time, one can easily observe that considerable sums are spared on OPEX, or, in other words, in expenses tied to its maintenance (since these tasks were never implemented). However, it is clear that by acting in such a fashion one has chosen to allow the asset's exposure to risk to grow considerably; if it is not maintained, the asset will of course be more likely to go through a failure or a dysfunction. But in the space between excessively lessened costs and increased risk-exposure, one can find zones of optimal profitability which we can now pinpoint efficiently. This is the very core of risk monetization.

Indeed, if one is able to assign monetized and quantitative values to the points of these two curves, one may thus obtain optimums that will in turn make it much easier to access a standard of fully informed decision-making. This fundamental principle is a reliable source of success for a vast majority of cases related to the operation and maintenance of assets (Figure 8.1).

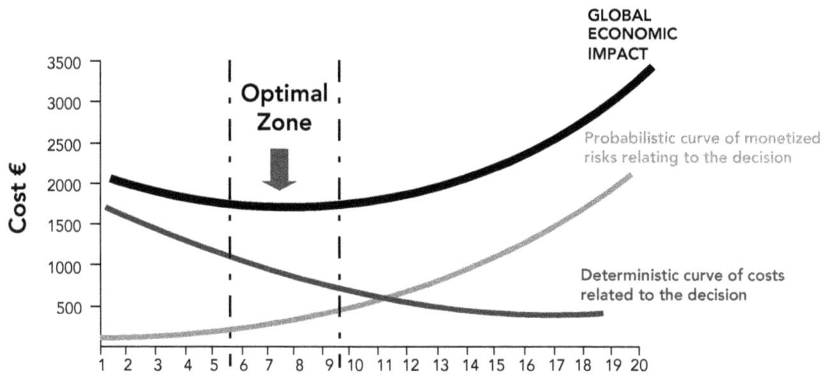

FIGURE 8.1 The risk/cost trade-off applied to operating assets

A number of industrial sectors have already incorporated these teachings and channeled them into effective internal processes to systemize their application on a day-to-day basis. As far as I can tell, there is no reason for the growing interest around this practice to drop in the foreseeable future. In a decade at most, technological progress in this field will undoubtedly allow us to access tools of risk monetization directly on our smartphone screens in real time, at the very instant a problem arises on the industrial field.

It is well established that the Asset Management journey's trajectory, which leads organizations on the path toward an optimal management of their asset fleet, is rooted in the phenomenon that we have just described, namely that of the monetization of risk in view of an optimum.

However, one must keep in mind that this scheme is not valid for every level of the Asset Management cycle. In such cases as that of the management of spare parts inventories, for example, the choice of an optimum is unable to guarantee a higher standard of decision-making.

The question therefore becomes: how does one articulate spare parts management with an Asset Management mentality geared towards the extraction of value from assets? Indeed, we will often struggle to evaluate demand. We might not know the volume of spare parts, and some plants don't even know how many parts are currently in use, yet they keep purchasing on criteria unrelated to the reality of demand or the volume of orders. Not to mention the multiple cases of parts deteriorating due to inadequate storage—when comes the time to install them, they prove to be out of order or liable to damage the machine. Under such conditions, managing a warehouse properly takes a miracle.

What I suggest is a brand-new approach to stock management that accounts for the global economic impact of stock policies and, generally, of the supply chain. The impact is nothing other than the sum of various types of costs and expenses: the purchase price of parts, the cost of managing the stocks, and the capital tied up (this is basic stuff). Add to this the economic impact of having a part unavailable that translates into a shortfall (while this is the most common impact, it isn't the only one). Thanks to this new approach, the equilibrium point between the expense incurred by having a part in stock and the risk of it being missing when required is easily found.

Indeed, Asset Management examples have successfully demonstrated that in this specific context, measuring the optimal risk/cost balance does not enable organizations to implement a pertinent management system for their spare parts inventories. This can easily be explained. It is clear that in this particular trade-off, risk is mainly embodied by the cost of the unavailability generated by the shortage of a part, and that the shortfalls are obviously widely superior to the CROs of the said part. It is therefore quite logical that in a risk/cost trade-off, depending on an optimum would only encourage industrials to buy too many spare parts. This observation is the practical base to the development of the concept of *consented risk*, which invites managers to disregard the optimal risk as long as they stay underneath the initial risk level—the level measured before proceeding to the optimization.

8.3 Consented Risk Optimization

Risk/cost optimizations do not necessarily have a positive global impact as far as extracting value from assets is concerned. This is what we mean by consented risk optimization. This innovative approach, developed by my team, is not only able to pinpoint the optimum policy regarding spare parts but also to highlight the risk exposure induced by straying away from the determined optimum (increase/decrease in risk against increase/decrease in the number of parts in stock). Today, this approach is obviously implemented via consented risk simulations of optimization through Assets Value.

The results found will ultimately serve to implement the recommendations in parts referencing, regarding volume, costs, and risks, immediately after they come in. For this reason, we will highlight the need for industrials to opt for a revision of their stock policy with these results and conclusions in mind.

In fact, regarding the parts references that require a variation in their stock volume in order to reach the proposed budget cuts, two paths are possible:

- Destocking the parts with the least risk attached (and their economic equivalent)
- Stocking parts that strongly reduce the exposure to risk (and their economic equivalent)

Various levels of consented risk can then be implemented gradually and sparingly, always keeping the final risk below its original level and accounting for every operational and administrative constraint (Figure 8.2).

EVOLUTION OF GLOBAL ECONOMIC IMPACT

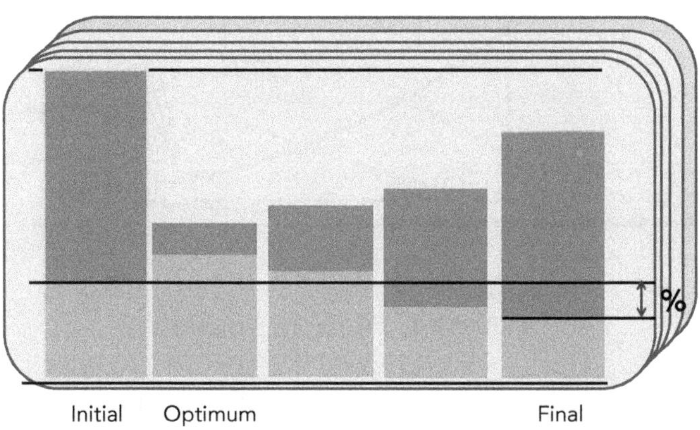

FIGURE 8.2 Global savings on stocks with consented risk

However, it is important to keep in mind that there are limits to the acceptable degree of risk, which can only go so far above the optimal level. This consented risk is *measured*, and is only consented to in light of its benign character. As we've already discussed, consented risk cannot exceed the level of risk that applied prior to the optimization. If an Asset Manager deems it necessary to encourage an operator to expose himself to a degree of risk, he can only do so within a strictly defined range.

To exposit these notions of consented and measured risk, one can refer to what I've come to discuss as the cyndinical rectangle. Cyndinics is quite precisely the science of risk: the cyndinical rectangle is a simulator-generated model that graphically illustrates the area of uncertainty contained in the risk/cost trade-off. In order to optimize this practice, the role of the Asset Manager must therefore be to strive to reduce this "area of uncertainty" in order to efficiently assist the decision-making process (Figure 8.3).

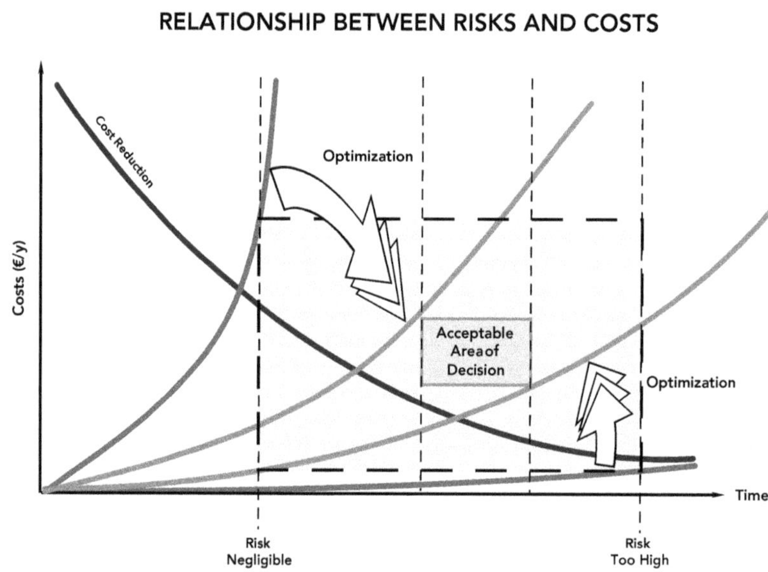

FIGURE 8.3 The cyndinic rectangle

The introduction of the concept of consented risk has brought to light previously unknown aspects to the notion of risk aversion. Let us consider the case of "traditional" asset operators: they neither calculate nor monetize their levels of exposure to risk, which has never prevented them from producing satisfyingly. It is truly surprising that the emergence of the monetization of risk, and specifically that of consented risk, would come as a surprise to industrials who seem to be discovering, through the recourse to graphic models, that they routinely consent to forms of risk. It is however quite evident that the modeling of risk does not create a new situation, but rather shines a new light on a situation to which these industrial managers had already consented, although perhaps unknowingly and therefore without discomfort. In fact, it can be observed that leadership tends to shy away from consciously encouraging practices that imply degrees of risk exposure, which brings us back to the idea of managers deploying non-committing policies in the face of economic risk. It is easier for decision makers to remain in ignorance regarding the vulnerability of their asset fleet and the shortfalls that this vulnerability could generate; yet we can quite clearly see how crucially useful it would be for these decision makers to be adequately aware of these factors in order for them to develop optimal management strategies.

8.4 The Language of Risk

We can easily distinguish between three broad families in the field of industrial risk:

- Operational risks (tied with the dysfunctions of the productive apparatus)
- Corporate risks (tied with the HSE factors and which also cover human behaviors within the organization, brand image, etc.)
- Financial risks (tied with industrial economics and sometimes the macro speculative scale)

Asset Management understandably focuses on the operational risks, since those are directly tied to the assets themselves. But our discipline does not neglect the management of corporate risks, which determine important factors such as staff and customer security or environmental preservation. Furthermore, it is not disjoint from the matter of financial risks: these two fields are brought together through the econometric translation of assets, or, in other words, through the measurement and the economic translation of technical risks.

To put it briefly, it is well established that risk in itself is not a homogeneous construct. It is rather strongly diverse, as risks distinguish themselves through their typologies, their sizing, or their consequences. This is indeed the primary source of legitimacy for risk monetization. In order to prove this point, one need only recourse to a *reduction ab absurdum* (or proof by contradiction): indeed, if risks were homogeneous and equal, then we would have been able to develop tools facilitating the daily management of trade-offs (thus eradicating operational risk) a long time ago. But since these risks are heterogeneous, we need to rely on experts who can specifically treat risks as they manifest themselves on particular occurrences. Following the industrial catastrophes in the chemical installations of Seveso (in Italy), a number of European guidelines clearly stated that industrial risks should not be considered via the same prism and expertise as managerial or financial risks.

One could express a form of shock when faced with the idea that these polymorphic risks are often reduced to a single economic dimension. But let us put things into perspective: Asset Management primarily focuses on the productive apparatus (infrastructures and production lines) and the idea of monetizing risk is only designed in a very targeted fashion, to patch a historical flaw that prevented decision makers from adjusting their choices other than through a corporate "no" or their own incapacity to decide. The monetization of risk does not imply the commodification of the human or environmental factors; it is simply a precious tool that allows us to base corporate decisions on facts rather than on oft-misleading impressions and feelings.

8.5 International Benchmarking: A Bittersweet Experience

Over time, the concept of benchmarking (which entails, as we will attempt to demonstrate, so much more than a mere "good idea" in terms of business capabilities) has become in some sectors of economic activity both a compass and a lantern in the hands of market analysts. More strikingly, industrial leaders and infrastructure managers have succeeded in appropriating the parenthood of this notion over the last two decades. The pertinence and usefulness of such a tool are therefore well established, but it would be misguided to assume that the track record of this practice is a fully "sweet" one: in fact, if one were to study the use that has derived from the emergence of benchmarking, one would be faced with a contrasted picture, which would undoubtedly reveal hints of bitterness.

I often make the assertion that benchmarking presents an effective analogy to the game of poker: indeed, in these two contexts, the main stake lies not in the "cards" that are dealt, but in the methods and stratagems deployed by the players. In order to illustrate this assertion, one could point out that anyone working in Asset Management or in another discipline whose members are regularly invited to participate in our international conferences and congresses will frequently have heard a speaker launching his presentation with the following claim: "Our organization is a benchmark throughout the world." Undoubtedly, benchmarking has allowed, through mass effect, for the rise of intrasectorial and intersectorial performance scales. But in parallel to this positive input, we cannot overlook the fact that this trend has induced and continues to induce in numerous cases a regrettably short-lived enthusiasm. Hence, to perpetuate our analogy, we could state that the "deal" is good in itself, but that the players must now master it in order to fully reap the potential profits. In some cases, we even hear managers giving such speeches as: "If such organization has been able to obtain such indicators, then we should obtain them as well." However, benchmarking is not an exact science. Worse yet, the data that feeds benchmarking is treated based on supposedly serious and often voluntary collections but it is rarely checked with the sources. When a brilliant presentation based on benchmarking-oriented numbers is given, one can therefore only trust in the speaker's good faith or hope that the numbers are extracted from a reliable polling institute.

In very simple terms, benchmarking is synonymous with comparison. Quite obviously, it is something to operate a comparison and another thing altogether to *use* this comparison without possessing enough tangible evidence regarding the indicators that we should wish to transpose. It seems important, then, to come back to the *finality* that lies at the very genesis of the concept of benchmarking.

In order to practice proper benchmarking, one must base one's decisions on evidence that is beyond doubt, available, based on recent information, and relative to functional and well-performing elements. These factors are essential and indivisible: the elements must be simultaneously indisputable, accessible, actual, and relate to a very satisfactory operation. If we were to put the information commonly presented as "benchmarking" to the test based on these requirements, how many could walk out "with their heads held high?"

It is therefore clear that benchmarking is a valid idea but that its foundations must be well understood and applied with a sense of rigor. It would be easy to demonstrate that this definition of benchmarking would make any comparative analysis much more efficient than that which merely transposes the enviable indicators, in order to set an installation's targets and objectives through a pure imitation of those initially set for another plant.

In other words, implementing benchmarking as we've defined it above implies practicing an activity that is radically distinct from a comparison of results. The principle is quite clear: benchmarking is not a comparison of results, indicators, numbers (or records!), but rather a comparison of the *mechanisms* through which superior performance becomes evident.

Indeed, benchmarking as it is practiced today pertains to imitating what is visible and apparent in terms of numbers, but which in the end often refers to the less important features: the more accessible data is compared to the detriment of the high-stake performances. Nonetheless, the "sweetness" (or the input) of benchmarking lies in the fact that it unquestionably provides organizations with quick wins in the shape of easy and swiftly tangible progress. This is a very positive aspect to benchmarking, since this practice encourages organizations to develop new internal experiments based on external rationales.

But aside from these quick wins, one must also mention a real bitterness linked with this practice. The main problem identified in benchmarking as it is practiced is that it is rarely guided by hard evidence and undisputable facts. Pr. Pfeiffer demonstrates, in his essay *Hard Facts, Dangerous Half-Truths And Total Nonsense: Profiting From Evidence-Based Management,* that only 15% of managerial decisions are truly based on evidence. According to the author, this trend encourages the multiplication of gurus who, by appealing to their customers' fantasies, have no need to resort to the use of tangible and solid evidence. Indeed, if gurus relied on scientific data, we would know by now; however, gurus typically rely on belief and faith rather than objective elements. This "bitterness" brings us to the perverse effect of benchmarking, commonly known as "informal benchmarking." This is a phenomenon that we could best describe as "emulation free from criticism," since it brings those who rely on indicators to try to generate improvement solely via the imitation of the alleged "best cases." Disregarding the lack of rigor that it embodies, this attitude is also dangerous due to the fact that it brings the corporate leaders to make emotional decisions.

In order to understand this notion more precisely, let us picture the following scene: in an international congress, the speaker for a certain activity sector presents his "benchmark," which makes a strong impression on his competitor, also in attendance. The latter returns to his office, fulminating with jealousy due to the strong performances described in the presentation, calls for an emergency meeting with his colleagues, and demands that within six months his organization should obtain the same indicators. This is a tangible example of a very widespread tendency: to draw inspiration from numbers and indicators and to strive to imitate the results exhibited by competitors, which causes professionals to turn a blind eye to the more pertinent factors of success, which should be at the very center of any rigorous comparison.

The prevalent situation of informal benchmarking throughout the industrial world can be attributed to many causes. To expose these reasons, we will begin by considering the elements that are used today in typical benchmarking projects instead of the more pertinent (but rarely available) data:

- Results/indicators of the competitors (usually published rather than obtained from the source)
- The period over which this information was calculated

But aside from this information, a number of factors that should be considered are often neglected:

- Evidence that the challenge faced by the professional whose indicator is being studied is identical to that faced by he who will "use" the indicator. The first criteria of success is this assurance that the comparison applies to similar or identical problems.
- The solution-building mechanism, and therefore the logical and methodological base for the obtaining of a good result, which should be the motivation for the comparison itself.
- The verification of actual chances for transposition.

Consequently, I would advise those who intend to practice benchmarking in the context of Asset Management to steer clear of five pitfalls which I've identified over the past few years. It is important to ensure that one does not:

- Base his comparison on obsolete data
- Base his comparison on the personal experience of one professional
- Allow oneself to be fooled by passing trends
- Allow oneself to be fooled by dogma and unfounded beliefs
- Practice the inconsiderate imitation of numbers

Let us conclude this intervention on a positive note, which shall undoubtedly reassure those who intend to apply benchmarking to Asset Management. If we operate a synthesis of the variety of disciplines that make up Asset Management and of the easily observable pitfalls in benchmarking practice, we can promptly understand that trying to obtain outstanding results by imitating those "of the competition" without any critical perspective may lead to informal benchmarks that turn out to be toxic for the organizations who implement them. Now more than ever, any benchmarking must require concrete evidence before giving way to any transposition. Accordingly, Asset Managers must accept a level of uncertainty to transpose high-evidence results. Hence, one should keep in mind that it is always more reliable for the Asset Management to trust in the "windshield" provided by a solid comparison rather than in the "side mirror" embodied by ideas that may look appealing but that are not guaranteed to induce a brilliant future.

CHAPTER 9

Operation and Maintenance

Discussing the operational stage in the industrial asset's life cycle brings us back to the very sources of the Asset Management genesis. Indeed, one must keep in mind that a good deal of the essential motivations for the Asset Management initiative in the 1980s arose from observations tied to the operation of industrial assets. Of course, this isn't to say that prior to Asset Management no one strived to make machines and infrastructures work more efficiently. Nor is it to say, for that matter, that prior to our discipline no one worked towards successfully implementing programs destined to improve the assets' functioning.

Following this rationale, one could wonder why the operational phase has been such a crucial vector in defining the problematics that would find their answers in the development of the Asset Management line of thought.

9.1 Of the Usefulness of Monetizing Risk

To treat this question correctly, one must come back to the central question, which is that of the relationship between industrial maintenance and Asset Management. As we've already noted, the task that consists of maintaining the condition of an asset is a noble one, but it is distinct from that which seeks to extract value from an asset throughout its life cycle. In other words, from the perspective of industrial managers (among the ranks of which I once stood), it seems that every progress that has emerged in the optimization of maintenance and production tasks in the past was praiseworthy but necessarily incomplete.

As one can recall, we've already presented the idea that for some time operational decision-making has seemingly been "walking on a single leg." In fact, operational managers are quite acutely aware of the fact that any decision they may make will have corresponding cost, on the short and the middle terms. What is less intuitively apparent to them is that when they elect *not* to decide, the impact of risk—a notion that they perceive as abstract and painless—naturally conducts them to the idea that what is unknown may not be managed and therefore creates that most comforting situation where no expense seems necessary. This is where Asset Management culture has operated the most fundamental paradigm shift among managers: it has made contradictory elements, which were previously regarded as inconsequential due to their uncertain nature, become available and intelligible.

Moreover, thanks to the development of *risk monetization*, the taboo has fallen. This has had a double impact on Asset Managers. On the one side, the monetization of risk allows them to make decisions in full awareness of what lies on "both sides of the scales," or, in other words, to base their decisions on objective and quantitative elements that describe the potential impact of the risk. On the other, the information obtained via the monetization of risk is boosted by various elements of empirical comparison, since we now possess hundreds of feedbacks from all over the world—another benefit of the outstanding factor of internal communication that Asset Management has become.

We could thus quote a number of key agents in the industrial sphere who have been able to take an important step forward in the development of their activity thanks to the innovative and concrete tools and methods promoted by Asset Management. It should therefore come as a relief that the operational and maintenance sectors are now equipped with such a potent arbitrational factor to solve the situations in which the contradictory decisions taken could not be supported by impartial and quantified elements regarding the impact of risk. Today, these sectors have at their disposal a vast number of decision scenarios that examine cost-related risks on a single plan of comparison and thus direct the final choice toward the optimal scenario for the general benefit of the organization.

However, these tools were not adopted by everybody overnight. Because it calls for a new behavior as well as new practices, big change takes time to sink in.

In this sector especially, and despite the progress of technique, professional training, and management methods, a maintenance engineer today is faced with the same basic problems as he was in the 1980s. We certainly have access to modern stock management methods, accounting software, outsourcing, and externalization techniques and the likes; sophisticated devices will readily diagnose the deterioration rate on a machine. Nevertheless, come the final quarter of the year, the plant manager will without fail require an X% cut in next year's maintenance budget from the maintenance manager, who will follow orders against his better judgement. This happens because he lacks the arguments to question the targets, and he lacks the tools to quantify the risk incurred by cutting costs. How did we get here? Undoubtedly, neither the specialist nor the company have adopted an economic and financial approach to risks.

My father was an accountant. Thanks to him, I realized fairly early that while companies count the money that they make, they fail to count the money that they fail to make. Today, I have learned enough to confidently state, with proof at hand, that any plant that fails to account for the shortfalls caused by its strategic misjudgments is bound to go bust sooner or later. The time for change has come at long last. In industry, there is no leeway for inconsiderate risks, ticking time bombs, or losses. The maintenance manager ought to refuse to lead a frustrating professional life in which fires constantly have to be put out and parts replaced without notice. He must instead learn to behave like an Asset Manager.

If machines could talk, they would complain about being mistreated with no benefit to their owner as far as Asset Management is concerned. Luckily, we do not have to wait until machines learn to talk to realize that they deserve better care in order to produce more over their lifetimes, with more value extracted from their use. We now have the tools to understand them, and consequently to change our behavior in terms of global performance.

What this brings to light is the idea that risk monetization was the missing vector in the process of making activities such as maintenance become true factors of value extraction rather than tasks regarded as merely cost inducing. How often has an operational manager found himself in a meeting with his board of directors where he is asked about the cost that a negative response to his demands would induce and forced to answer, with all the discomfort that such a reply would induce in a skilled engineer, that he simply *did not know*?

Furthermore, a vast number of activities have profited immensely from the monetization of risk. Among these, we could name fields such as the optimization of the periodicity of intervals for maintenance, the prevision-making for the degradation kinetics of pressurized assets, the optimization of the planning of major shutdowns, the reduction of spare parts inventories with a consented risk, and many other elements that could never have been objectively quantified had we continued not to monetize risk, and therefore not to calculate trade-offs.

9.2 The Role of Trade-Offs in Asset Management

One of the major benefits reaped through this technological revolution in the field of Asset Management has undoubtedly been the rapprochement it has induced between the technical, accounting, and financial spheres. Indeed, in the past and unfortunately even to this day in a vast number of organizations, a form of structural incoherence has tended to complicate and make fragile the circulation of information within the organization itself.

Let us consider a very simple scenario to illustrate this notion. Imagine that you should wish to ask operational agents (or "field agents") to provide you with a number of cost-related elements or economic components that could be of assistance to your process of technical decision-making. These agents would fatally refer you to the accounting department, which is oft-regarded as the exclusive authority in the field of numerical values. But should you address your request to the accountants, they would soon reply with that most enigmatic phrase: "We are here to count." Well, they certainly are, but *what* is it precisely that they're meant to be counting? "We are here to count the data that we are sent." Quite so, but can they question or analyze this data they receive, in order to help you obtain the information that may help feed your process of decision-making with objective elements? "Of course not! Haven't you heard? *We are here to count.*"

It is clear that this scenario is caricatural, and cannot be applied to the case of every particular capital-intensive organization. However, we should keep in mind that organizational structures have always tended to be partitioned and opaque, to the extent where crucial information could not freely circulate from one sector to the other. In this regard also, the synchronization of objectives due to policies of alignment marks a shift in the functioning of industrial organizations.

Hence, new Asset Management Systems are in a much more advantageous situation than they were historically. But they are also faced with a challenge: in order to be successful, a system must implement a route that ultimately leads to a strict correspondence between Asset Register and Asset Costing. Therefore, one of the priorities of contemporary Asset Management should be to methodically ensure that a one-to-one correspondence is established between the register of a physical asset and its equivalent in the asset accounting base within each organization's informatic systems.

To implement the alignment of Asset Register and Asset Costing, one must provide decision makers with equivalent information for these two dimensions of the decisional process. Consequently, this mass of information will allow managers to develop scenarios based on objective data, which will therefore be more likely to be successful scenarios. This ideal situation for decision makers is unprecedented in the history of industrial organizations and of their operational staff.

It is the spread of the usage of trade-offs that has made this new availability of crucial information possible. As we've said, a trade-off is fundamentally nothing more than the calculation of an optimal balance for the relationship between two typically contradictory variables. The phrase has actually gone on to acquire specific meanings in various contexts. In our sector, the most pertinent synonym would probably be "arbitrage." A number of distinct trade-offs are commonly used in Asset Management. The most frequent, in the operational phase, is that which determines the relationship between the risks and the costs tied to the assets.

The risk/cost trade-off is a most efficient tool in the resolution of the operational issues that we've discussed throughout this text. Before reaching the chapter in which we will try to shine a light on the widely successful applications of these techniques, it is important that we insist on the pertinence of the use of this trade-off in the operational and maintenance phases. It hinges on two factors that must be considered when making decisions about industrial assets: the management of uncertainty and the distinction between the useful life and the mature life of an asset.

9.3 Managing Uncertainty

If machines could talk, this is what they would say about our habit of basing decisions on historical data:

> *"The eight machines that make up the production line of this plant are almost identical. The only difference between us is our age: some of them were already here when I arrived; others came later.*
>
> *"I find it intriguing that all of us receive the same treatment regardless of our age. The machine next to me is the oldest in the plant, and she creaks all day long. She is increasingly tired; she deserves more care than I do, for I am in the prime of life, fully fit, and she is not. In fact, the maintenance technician applies the same treatment to all of us whenever he visits.*
>
> *"I have always found it very strange, and it was only the other day that I understood the how and the why. In a conversation that I overheard, the maintenance manager was explaining to the new intern that the boss uses the history of each of our past failures to draft the maintenance plan of the machines. So, this is how you allocate the care that each of us receives: you look at our past failures to prevent the ones to come? It's crazy! If I hadn't heard it from the boss himself, I wouldn't believe it.*
>
> *"All of you that work in maintenance should know that our failure history says nothing about our present or our future state. My neighbor might not have suffered many failures in recent years, but the fact is that she is exhausted. She has more and more problems every day, and you can't even see it! Poor soul. What are you waiting for before you take care of her? Until your history (hers and others) shows you that she is currently failing too often? This must be it. I ask you to transfer some of the maintenance tasks that were planned for me onto her, as a favor. I am in the prime of my active life. I am the fittest and the most powerful in this plant, and I don't need that much looking after. Let me guess: you use the average failure rate of all the machines to determine how much care I need, is that it? How humiliating!"*

Let me tell you: you have no grasp on the reality of this plant. Stop checking the rearview mirror and try planning for the future by looking out of the windshield.

Undeniably, 99.9% of maintenance professionals base most of their decisions, actions, policies, and strategies on the failure history of their machines. Planning preventive maintenance this way is widespread in this sector. Those who work in new industries, with brand new machines that don't yet have a history, will even inquire about this precious history to inform their maintenance practices from people who have used the same machines for longer. This seems like a sound enough approach since all of us have a technical memory that informs our decisions, intuitively or rationally. Basing our daily decisions on past experiences, be it our own or others, makes sense.

Although established and widespread, this practice has limits. Considering the knowledge and means available to us, it is not the best option. I realize that some of my colleagues will call me blasphemous, yet my long experience in the field supports this conviction. In this chapter, I will thus propose a new approach to preventive maintenance planning, based not on the formal history of past failures but on their tendency to occur in the future. I call it "managing uncertainty:" it makes the job of the failure's analyst quicker and more precise, and it also helps to calculate the MTBF based on propensities (data deduced from the curve), without reverting to the failure history (curve deduced from the data). This idea builds upon the argument for the demystification of the MTBF already introduced in Section 7.2

The first and the best argument for "uncertainty management" lies in the margin of error attached to any failure history. Comparative studies made on various CMMS systems (Computerized Maintenance Management System) prove it: in any history, at least 35% of the data is incorrect. The margin of error never drops below this number and cannot be reduced.

This will come as no surprise to those of you who work in maintenance. Like most professionals in the field, you had an inkling that much of the information in your database was unreliable. Note that this isn't specific to your company: it is a well-known fact, backed up by precise statistics. But the question remains: why is it the case? Because the data is collected by a human employee. Once more, the often discussed "human factor" rears its ugly head.

In the early CMMS, the maintenance staff would input filled index cards into the failure history. We know that employees would generally conceal the truth in their filings for fear or shame of revealing their poor spelling. Everybody knows that technical data needs to be precisely drafted to avoid distorting the information—and thus the derived statistic. While this is not the only explanation for the inaccuracies in histories, it is certainly the major one.

Obviously, CMMS has evolved past the obsolete method of inputting failures by hand. Some use the convenient barcode system; one only has to look into the catalog, localize the barcode that matches the identified failure and scan it. After ten years of running with this system, we realized that the margin of error was only slightly smaller than with the manual system. Why? Because man is unfit to read barcodes, and, in the process of scanning, he tends to get them mixed up. On top of that, imagine that a colleague shows up to discuss the soccer game, or calls him up for lunch, or any other distraction. You can guess the results.

Today, data collection and reading are made easier by pocket-sized electronic devices, tablets, and digital mobile solutions. These are directly plugged into the failure database. This process is expected to further reduce the margin of error, but it is still early to draw any definitive conclusions.

In fact, CMMS are consistently fine-tuning, becoming more reliable and widening the range of possibilities that they offer. Today, the range of software dedicated to failure history management is large: these wonderful programs will draw any types of curves and graphs to present the information in the desired fashion. Maintenance professionals spend an increasing amount of time on these systems, gathering past information to inform their decisions about the future of their machines—and this is the reason for my second argument in favor of a new approach to preventive maintenance planning. Chapter 7 highlighted that making decisions based on the failure history was akin to driving with both eyes on the rearview mirror. As important as the rearview mirror is, looking through it will not tell you which way to go: one has to look forward! What you see through the rearview mirror is not enough to drive a car—one has to look through the windshield (Chapter 7—"Operational Reliability").

The past behavior of a machine is not necessarily correlated to its future behavior. Its operational mode, its natural wear or any such factor could dictate a tendency that contradicts the scenario outlined above. Another crucial factor will be analyzed in detail in the following chapter: the machine's lifespan. Statistics show that every machine is more prone to failure toward the beginning of its life cycle. They stabilize and become more reliable after a while, with failures occurring occasionally. Then, towards the end of their life cycle, the probability of breakdown increases. A machine behaves similarly to a man. Health problems run the chance of being many at birth (see child mortality); his health will be stable during youth and adulthood; as he ages, his health deteriorates until death. The "bathtub curve," devised to model the biological probabilities of natural death, here transposes to industry: the occurrences of machine failures are modelized over their lifespan.

Furthermore, the analyst will generally base his conduct on the average number of failures within a given group of machines. This scenario is comparable to the one outlined at the beginning of the chapter: the machines with the most damage receive the same treatment as the undamaged ones. Is this a sound basis to formulate an efficient preventive maintenance plan and avoid future breakdowns? I doubt it.

Dear reader, dear industry worker, as fathers and brothers, let me ask you the following question: to what extent was the history of your oldest child useful in understanding and educating your youngest child? If you considered it to have been 100% useful, please write to me: I will seriously investigate your case!

This brings us to my third argument: the raison d'être of any maintenance professional is not to keep count of occurring failures. It is to avoid them, or at least limit their consequences, since they cannot be fully avoided (just like sickness and death cannot be avoided in a human life). At best, the "zero failure" dogma is a fantasy; at worst, it could be a serious strategic and economic mistake.

By overstressing the importance of failure histories, the maintenance professional wastes precious energy in inputting failures that should have been avoided in the first place. Imagine if, as I wrote this book, I spent more time counting my spelling mistakes than I did trying to avoid them. It wouldn't make sense, would it? In my opinion, this is exactly what maintenance professionals do with their failure histories.

Ironically, the data so painfully collected ends up stuffing up databases so complex that they lose all practical usage. Like the chests we keep in the storing cabinet: they are so full that, after a while, it becomes impossible to find the item that we are looking for. On an international scale, CMMS record 95% of the failures that occur in plants, only 30% of which are analyzed. Less than 4% will eventually be used to inform a decision. Isn't it incredible? To pick up the analogy: there is definitively a box of screws hidden somewhere in our chest, but it is so hard to find that driving out to the hardware store for a new one is actually easier.

Some of my readers might be thinking: "Still, we need objective data to work with, and the failure history is our only source of information." You are right in a way. You do need data to work with. However, you won't find it where you have so far been looking. This reminds me of an old joke: a man comes home to find his wife on all fours in the living room. When he asks why, she claims to be looking for her wedding ring that went missing in the bedroom. "If you lost it in the bedroom, why are you looking for it in the living room?" he asks. She replies: "Because there is plenty of light here!"

Let us keep in mind that the most relevant data to preventive maintenance planning would be the machine's level of damage, not its level of failure! We could then deduce its tendency to break down and anticipate the date. Since damages are tough to measure, the maintenance professional tends to go "where there is plenty of light"—the failure history. There is "light" enough for everyone to notice and count the machine's failures. Our databases are full of fantastic failures, beautiful and over-complete reports. This is how the man whose job is to prevent failures spends his time adding them up. I've sat in countless meetings in which the CEO was presented with beautiful graphs, filled with months of collected data and showing tens of failures that demonstrate the poor state of maintenance. The CEO would then raise the tone on the maintenance and production managers: "What you are telling me is that it took all of these failures for you to realize the true state of maintenance?" Let us be reminded that anticipation is a key quality for managers.

Let me be quite clear: I am not against failure history. If you see me tomorrow, giving a class or a conference, please don't act as if a sign on my forehead read: "Down with failure history!" Collecting the data is one thing; knowing how to use it best is a different thing. Undeniably, such histories do help to improve the quality of failure diagnoses. It is therefore an efficient way to enhance the efficiency of maintenance, and thus of production. Nevertheless, a history of damages would be better suited to other aspects of Asset Management (such as determining a preventive maintenance program and purchasing spare parts). Unfortunately, machines can't talk. They can't communicate what they feel and where they hurt. If only they had that incredible damage recording system that humans enjoy and that we call pain! But the machine doesn't share its pain; it only ever shows anything when it breaks down. The fabulous progress made in conditional maintenance and preventive tools—infrared cameras, vibration recorders, tribology tools (that measure friction)—does not cover the huge variety in types and modes of failure.

You might wonder: "If failure histories are too thin to plan maintenance on, and if damage histories are an impossible undertaking, what options do I have left? How will I calculate the MTBF, which, as part of the history, is a key aspect of my job? What is to be done?" The way out is to tackle the problem from a different angle. It is not the only angle, and it is not meant to turn into a dogma. It is simply a new approach that is spreading across the profession. Regardless of whether you choose to adopt it, I would be content if you accept the reservations that I have so far expressed, in the interest of demystifying the maintenance policy that professes to use the damage history as a starting point and turns it into the only tool used to calculate the MTBF.

Failure histories are done in an effort to pinpoint a machine's degree of criticality, that is, to what extent and with what impact it tends to break down. Let us use a circle as a metaphor for the criticality of a machine, supposing that an increase in the diameter of the circle indicates an increase in the degree of criticality.

Consider Figure 9.1: When using data history as a basis, it might take many years of recording before enough data (failures) is gathered to induce the machine's degree of criticality—that is, the size of the circle.

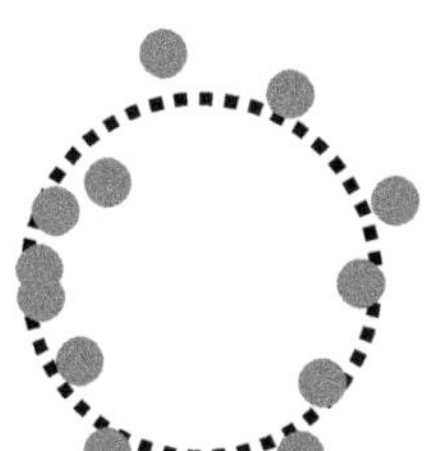

FIGURE 9.1 Analogy between a circle's diameter and a machine's degree of criticality as deduced from its failure history

On the other hand, my approach—uncertainty management—does not bother with breakdowns and the rearview mirror; instead, it stares straight ahead, projecting a tendency and anticipating. The first step is to enquire, from the maintenance technicians that interact with the machine daily, about the chances of breakdown. Being well aware of the state of their machines, and having tracked their breakdown history, the answer they provide will be informed by their professional experience and their own perception of the machines.

In geometry, three dots hanging in space are enough to draw a circle. The right questions, directed at the right person, the right way and at the right time, will provide the three dots required to know the size of the circle. Isn't it interesting? This operation, known as elicitation, relates to the way one extracts information from an expert. Elicitation shows that a small amount of data (contrary to histories) and dots is actually necessary; when drawing a circle, three dots suffice (Figure 9.2).

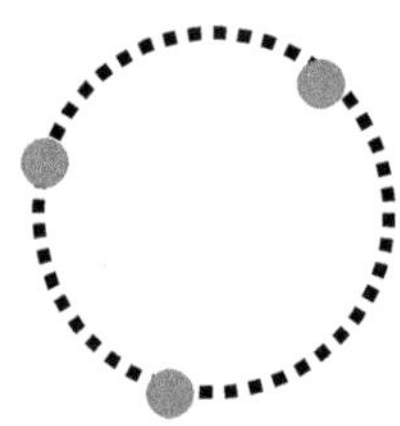

FIGURE 9.2 Circle drawn from three dots

One might argue that, the more dots, the more precise the circle will be. But what we are concerned with here is uncertainty: we know that estimating a machine's next failure based on its failure history is as uncertain as using the maintenance technician's opinion as a base. Nevertheless, the latter approach has two benefits. It will buy you time: the process will give you a fairly accurate overview of the machine's state much faster than if you collected its failure history over months or years. Furthermore, it will be based on what is currently happening to the machine as opposed to what has happened in the past. Its everyday environment is better accounted for.

In order to ponder the degree of uncertainty (small yet real) of this approach, a range of experts ought to be questioned. They will have diverging opinions on the machine's criticality; the same reality will be read optimistically by some, pessimistically by others, and neutrally by the rest. If three technicians provide different answers, three different circles appear (Figure 9.3).

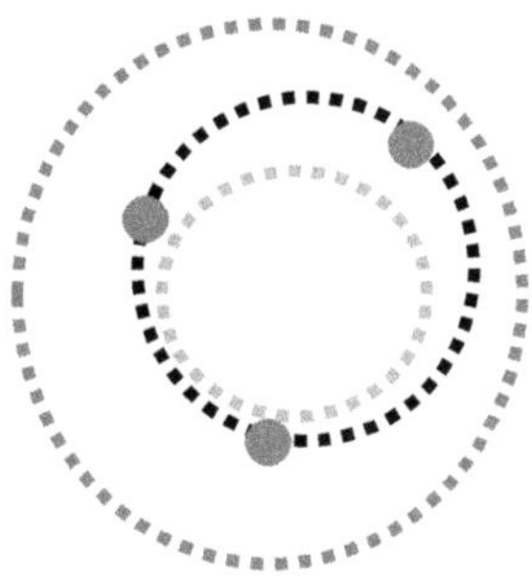

FIGURE 9.3 Looking for extreme scenarios through an uncertain estimation

The next step is to model the information provided by the technicians in order to establish the risk curve (what are the risks incurred by those preventive maintenance tasks over a given time period) and the cost curve (the monthly cost of this task over the machine's life cycle). This will be further illustrated in Section 10.4—risk/cost trade-off and "brick" analysis of decision-making assistance in Asset Management.

On the curve that depicts the global economic impact, determine the optimal point. Repeating the operation for our three scenarios—optimistic, pessimistic, and neutral—will result in three different curves to depict the impact. Now, paste all three curves in a single graph and demarcate the interval between the optimal point on the pessimistic curve and the optimal point on the optimistic curve: the space created is known as the maximum range of decision (Figure 9.4). Any points outside of this space can be dismissed at once, since they will incur high costs and risks. The best possible decisions belong within the maximum range window. Thus, the manager will only have to opt for the decision that most reduces costs and risks.

For decades, the "bathtub curve" has been a crucial tool of decision support when considering the behavior of assets in the operational and maintenance phases (Figure 9.5). However, we should be careful to debunk some of the myths that surround it, both in terms of its usage and of the concept of reliability that it brings into play. The most prevalent among these myths is that which brings operational managers to imagine that when an industrial asset is acquired, its inherent behavior as designed by the manufacturer will naturally and immediately follow a degradation kinetic that would be presentable as a bathtub curve. This is untrue, and if one wished to resort to this analogy, he should keep in mind that operational managers do not buy "bathtubs," but rather "cups."

9.4 Useful Life and Mature Life

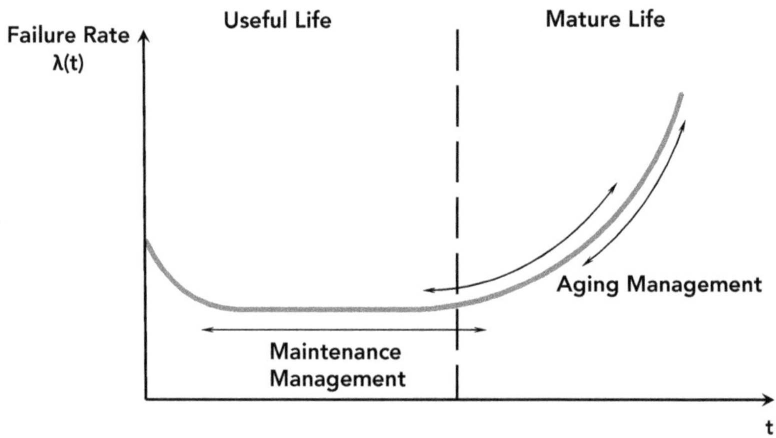

FIGURE 9.4 The useful and mature life of assets

HISTORICAL BELIEF OF OPERATION AND MAINTENANCE

In operational reality, the "bathtub" curve is not intrinsic to a new asset. In fact, it is gradually transformed into a "bath" according to the maintenance practiced.

FIGURE 9.5 The resulting "bathtub"

In other words, if a new piece of equipment has completed its burn-in phase (during which its reliability is in constant increase) without an efficient organizational maintenance strategy, the curve of instantaneous failure rate will invariably begin to show a rising failure rate. What we refer to as the "instantaneous failure rate" describes the conditional probability of witnessing a failure at a "T" instant while being aware of the percentage of surviving equipment from the same family of assets at this instant.

However, thanks to the effort provided by maintenance and on the condition that one is managing an efficient operational installation, one will always be able to bring the asset back to a lower intrinsic failure level each time that this curve will begin to rise. The curve being a stubborn one, it will tend to rise again immediately after the intervention led by maintenance, and once more maintenance will be tasked with bringing it back to its intrinsic standard. We are therefore faced with a rationale reminiscent of the famous principle of Hooke's law, which states that the extension of a spring will induce its return to its initial position. According to this principle, as long as the extension generates a return to the initial state, we can assert that the spring is in its elastic phase of function.

One can therefore experiment for oneself to bring the spring in such a state of extension that a return to its initial position is no longer conceivable: this tipping point marks the passage from the elastic phase of the spring to its plastic phase. In our general case that focuses on the bathtub curve, the analogy is almost perfect: hence, any industrial asset—so long as maintenance, thanks to its pertinent and efficient intervention, will allow the bathtub curve to return to the asset's intrinsic level of failure (or almost)—can be regarded as in its elastic phase relative to its degradation kinetic.

Fortunately, in the case of infrastructures and capital-intensive industries, the period during which most assets remain in their elastic phase is a long one. This period will henceforth be described as the useful life of industrial assets.

Reciprocally, from the instant when maintenance strategies can no longer impact the kinetic degradation sufficiently to bring the asset to its intrinsic reliability level, a growing failure rate curve appears—and this time, it will not stop growing. To stick with our Hooke-inspired analogy, this point marks the tipping from the asset's elastic phase to its plastic phase. In the perspective of Asset Management, this phase is known as the mature life of industrial assets.

This principle is a major indicator in the Asset Management mindset, in so much as the totality of the feedback that will come in the following paragraphs (related to the use of trade-offs or to the vast experiments in the application of Asset Management methods within industrial organizations), because it is focused on the operational phase, and can only describe the case of assets in their useful lives. The reader will understand through our presentation that the gap in Asset Management treatment between equipment in its useful and its mature life leads us to operate a necessary distinction between the management of the operational phase on the one hand and that of assets in their end-of-life on the other.

PART IV

The End-of-Life

CHAPTER 10

Asset Modification and Maintenance

As a fitting introduction to this chapter, which will strive to bring into light the attitude of Asset Managers during the second "phase" of the operational life of industrial equipment, it is essential that we come back to a phenomenon emblematic of the misdoings of industrial maintenance, already addressed in Chapter 7, "Operational Reliability," in which this notion was applied in relation to industrial assets during their useful life. I mean, of course, the loving care—this time in relation to assets at the end of their life cycle.

10.1 Loving Care: The Stakes of Therapeutic Obstinacy

If machines could talk, this is what they would say about the way industrial assets are modified:

> *"I know what you have in mind when I see you lurking around me, asking questions, standing around and scratching your head: you want to change something in me, don't you? Well, thank you for your interest in my efficiency, but please think twice before you try anything! Let's be honest: most of the modifications you've attempted did not prove successful. The latest one was a disaster. I might have gained a little productivity, but I have never been so prone to failure!*
>
> *"Let me ask you something: Do you consult with a third party before you modify me? My designer, or my manufacturer, perhaps? Oh, please*

don't be offended. I have no intention of questioning your methods, if only because you have proved to be a good professional and you've put a lot of effort into making this plant work. I only mean to recommend that you consult the people who know me best: this might increase your chances of success whenever you operate on me. Or you might operate differently. Or, who knows, you might even save money. Obviously, I would benefit as well, since I don't enjoy it one bit when your modifications fail. I am not a lab rat.

"On top of that, I know full well that all the modifications you experiment with are paid for by maintenance, isn't that right? This means that you pour the money designed for maintenance into transforming us. Whenever I have a part replaced, the people from on high will come here for a stroll and make this mean comment about me: 'So this is the machine that's been digging a hole in the budget and who still can't work right despite all of our modifications.' Now is the time to make a choice, old friend: Is your job to provide maintenance or to do 'additionance'?!"

Could any machine park go six months without a single modification? No. Undoubtedly, interventions are always necessary to enhance productivity or reliability. Therefore, this chapter will not question "whether" such interventions are necessary but rather "how" to go about them, since my job as a consultant has led me to visit countless companies in which modifications had gone wrong. The other issue I would like to bring up regards treating these interventions as investment (CAPEX) instead of as operational costs, as is more commonly done.

The underlying argument for any modification is that a machine will always fully comply with its attributed function. However, this isn't always the case for various reasons. In the first place, a machine might be attributed a function within the company's operational chain that doesn't conform to the specifications outlined by its manufacturer. Why? Because the manufacturer may not have described the machine's specifications 100% efficiently; the buyer may not have described his order 100% efficiently; between the salesman and the production team, the order specifications may not have been transmitted 100% efficiently; or the production team may not manufacture what has been ordered 100% efficiently. Furthermore, a defect could occur during transit or storage, and the technician might not set up the machine 100% efficiently. What are we left with? A machine that does not comply with what was initially decided by the engineers and approved by the industrial director. The machine's shortages will be noticed soon enough, and there will be no point in explaining the chain of events to the director, even less so to point out that what should work on paper doesn't always translate in practice, or to wave the bathtub curve in his face and point to the inevitable burn-in period. The director has purchased the machine for its written description: he expects it to work the way the leaflet said it would. Therefore, reverting to modifications becomes necessary.

There is a second reason for modifying a machine: to fix the damage caused by premature aging and return the machine to its intrinsic reliability level—or, at least, to an acceptable level of operational reliability. Modifications will also often be attempted in a bid to improve the quality and performance levels of the machine—that is, modifying its DNA. Another target might be to modify its operational mode—be it for safety reasons, to save energy, to protect the environment, etc.

Obviously, maintenance is entrusted with the task of modifying the machines. This is primarily because maintenance is generally thought to have a double mission: to operate the assets and maintain them in an operational state—not to mention the issues that might arise before a machine is started, which we have already discussed. Secondly, because maintenance is thought to be "the handyman." Thus, if plastic damages impede the machine from reaching its intrinsic reliability level, or if the production team targets record-breaking productivity rates, it will ask maintenance to "figure something out." The maintenance technician will take responsibility for it, and, lacking perspective on the plant's common problems, he will feel guilty ("I cannot bear to have a machine broken down!") and try to go at it alone.

This "let me take care of it" attitude is the root of most failures in modification. In an attempt to fix everything himself, the maintenance technician will usually opt to replace machine parts (or even its technology), which, as an occasional and timely solution, does not account for the machine's global reality, or its environment. While he might proceed with the best intentions and invoke rational arguments, the modification could nonetheless result in a failure.

To illustrate this point, let me relate what happened a few years ago with a public transport company that operates a subway system. Back then, subway doors were basically mounted with a coaxial, double cylinder compressed air engine and a three-way valve. The compressor would pump air into the valve to command the opening and closing of the doors.

But, after a while, these mechanisms tend to malfunction and get stuck, which impeded the movement of the doors; because the doors did not open or close properly, the subway would run late and upset the whole network. The root cause was discovered to be the valves: they had little tolerance for the humidity contained in the compressed air. They would corrode, malfunction, and get stuck.

Not only did it cause delays that impeded the efficiency of the subway, but it also threatened the safety of the passengers. One of the maintenance technicians thus took it upon himself to prospect for valves better equipped to fight corrosion. At the end of a long quest, he settled on two models, requested samples from the manufacturers and installed them on various subway cars. For weeks, the cars ran on empty around the warehouse; after a while, it became clear that one of the models was much more resistant to corrosion.

Mission accomplished! Naturally, the maintenance manager proceeded to order four hundred valves (on the maintenance budget) and had them installed immediately. He did so of his own accord, taking no second opinion, in the "I'll take care of it" spirit. Then, one Thursday morning, the entire city almost broke down. The valves might have been resistant to corrosion, but they were susceptible to the vibrations of the tracks (they were not conceived for this function). They were set up in cars that speed, brake, spin, and stop without notice, until they gave their last breath. They didn't last long and broke down in the middle of rush hour. In the reliability jargon, we call this a "change in the profile of the mission," which is a classic event in the life of industrial systems. The new valve might have been technologically and functionally superior to the old one, but it was not tailor-made to the operational characteristics of a subway. That day, the trains that had been fitted with the new valves broke down almost simultaneously, which caused an "underground blackout" on the line. Luckily, only a handful of trains had been fitted with them. The order was later cancelled.

You might think I'm exaggerating, but please note that I have too much respect for industry professionals to be ironic behind their backs. These incidents will happen if maintenance professionals insist on doing everything alone, notwithstanding that their attitude is legitimate and springs from the best intentions. In the example I mentioned, the technician grabbed the problem at the root—air distribution in the valve—and went looking for valves that could withstand the air distribution efficiently. Unfortunately, he overlooked a detail that the manufacturer or the company's engineering department probably would have caught. As such, keeping his intentions to himself was a mistake.

On top of the sense of guilt and the intuition to do everything himself, the maintenance professional is conditioned by a third factor: the loving care, his profound affection for the machine. The fact is that he loves disassembling and replacing parts. For this reason, in industry, modifications are done way more often than is actually necessary. This incurs expenses in organization, administration, purchasing, studies, interventions, and downtimes, and for what? For unsatisfactory results.

Let me be quite clear: I am not against modifications. I will readily admit that they are key to the smooth operation of a machine park. However, I am skeptical of their implementation, which is usually guided by pride, vanity, and loving care, and done without perspective. Before any decision is made, the production and maintenance teams should gather around a table with a few other people—the manufacturer, the designer, or anyone who has a direct relation to the machine. Successful modifications always result from collective and functional thinking: the systems vision so dear to Asset Management.

What do we mean when we speak of "*loving care*?"

This expression describes the attitude of consented blindness affected by the maintenance manager, who pretends to ignore that some of the degradations that have impacted a machine may never be treated because of their *plastic*—as opposed to *elastic*—nature. From that point on, the maintenance agent will ceaselessly fiddle with the machine, trying to ensure that it can follow the increase in its capacity dictated, for example, by market opportunity, and will only succeed in increasing the risks of damaging the piece of equipment. This is a serious paradox, as we are intuitively prone to assume that an excess in caretaking could benefit the machine's health. Yet, this is a wrong assumption.

In practice, reactive maintenance strategies (implementing the frequent dismantling and repair of machines), which are developed to maintain the machine's level of reliability, tend to have the opposite impact, actively facilitating the drop of its reliability level. Indeed, as the manager stubbornly tries to repair a machine against all logic, the costs become exponentially disproportionate: reactive maintenance necessarily implies downtime, which in turns generates considerable shortfalls. In this regard, we could say that *loving care* is also an accounting and financial pitfall.

Furthermore, this phenomenon is complete with a psychological dimension, which relies on the pride and "honor" of the maintenance manager. This justifies our recourse to the phrase "consented blindness." From the perspective of maintenance managers, it is *unacceptable* that the machine that they've been made responsible for should undergo degradation—and yet, it is precisely *unavoidable*. Thus, maintenance managers task themselves with an almost sacred mission: they work overtime and are relentless, refusing the ineluctability of their protégé's deterioration. This is closely linked with the "psychology of death," inherent to the medical world, that I associated with our discipline in the first chapter of this book; our job, as Asset Managers, is to accept that machines are programmed to decay as early as in their design phase.

When pushed to its excess, *loving care* implies anticipating potential failures and engaging in preventive action in the face of the expected catastrophe. But this form of anticipation is by no means as objective or "quantified" as the calculation of intrinsic reliability should be. It depends on *affects* rather than on rational elements. Anticipation *as such* is praiseworthy: it is indeed the very essence of preventive maintenance. But *affect-informed* anticipation, motivated by managerial *loving care*, is both costly and pointless and should therefore be proscribed.

This situation is rooted in a particular context that I'd described in my previous book as "the double failure spiral." On the one hand, leadership demands the impossible from the operational staff, as they wish to boost the productive capacity of the machines without inducing a drop in reliability; on the other, the operators themselves play along to the corporate game, which results in irrational affects and overzealous behaviors that may in turn only have a regrettable impact on the global economy of the assets and of the organization.

Above all, one should keep in mind that this situation is by no means a fatality. In reaction to this "double failure spiral," we must promote the urgency for the development of a "virtuous circle of success." This would take the form of an increase in dialog between management and maintenance agents as well as of a redefining of the task assigned to maintenance managers, which would thus cease to function as "fancy parts-exchangers" in order to fulfill the goal that should have been theirs. Theycould finally have a lucid and objective perspective over the condition of the machines whose optimal function they must ensure, and only act on the basis of their observations rather than be motivated by *intuition* or pride. And indeed, who could do a better job of pinpointing the risk/cost trade-offs than maintenance managers, whose activity demands that they operate on the frontlines of the assets' failures and be intimately aware of their cycles of degradation?

10.2 End-of-Life Threshold: A New Indicator for Profitable Operations

We've already spoken of the trend that consists of presenting the fields of Asset Management and industrial maintenance as conflictual or conflicting. We will spend more time discussing these allegations, but will also give some examples of key possibilities for improvement in the daily labor of maintenance professionals informed by the practical and conceptual input of Asset Management. We will therefore discuss the necessary discernment—between contradictory parameters on the one hand and separate realities on the other—that maintenance managers must acquire in order to improve their standards of decision-making.

In terms of contradictory parameters, we will come back to the notion of trade-off and particularly on the most useful of these trade-offs in the maintenance phase, that which sets out to evaluate the relationship between risk and cost. However, to discuss this topic in a coherent fashion, it seems crucial to begin by guiding maintenance engineers towards a more pertinent application of trade-offs approaches in order to ensure that they focus "on the right targets."

In more prosaic terms, let us consider the case of a machine in its mature life, which is very demanding in terms of maintenance labor but over which all attempts at repair are already relatively ineffective because of its obsolescence. As we know, it is highly unlikely that reactive actions could ever return this asset to its intrinsic level of reliability. It would therefore be pointless for a maintenance agent to proceed, as is often the case, to try and "maintain" this specific asset regardless of the expenses. We will not discuss in any more detail the corporate pressure that perpetually demands that failing machines work anew, nor of the specific pride tied with the maintenance line of work, which brings maintenance agents to refuse to accept the failure of a repair action for fear of coming off as incompetent. Our goal here is to provide operators and maintenance professionals with the necessary tools to orientate their maintenance strategies based on the character (efficient or not) of the act of maintenance itself.

In other words, why should we accept that so many technical interventions be conducted without proceeding to preliminary analyses that would allow us to determine whether the asset is still in its useful life (and therefore repairable) or already well into its mature life (which would make its repair infinitely more complex)?

Once again, being able to distinguish between the *useful* and *mature* lives of the assets appears a necessary step toward an adequate system of Asset Management (see Figure 9.4—Section 9.3).

As a reminder, let us state once again that a machine is considered as being "in its useful life" so long as it is possible to bring its (necessarily growing) failure rate back to its prior level, regardless of whether this level corresponds to the asset's intrinsic reliability. On the contrary, if interventions are no longer enough to temper a growing failure rate, one is entitled to assume that the piece of equipment has gone from its elastic phase of functioning to its plastic phase. Admittedly, the practical border between the ending of useful life and the beginning of the mature life, as witnessed in real life by operational managers, is not as strict and definite as the border between elasticity and plasticity would be. Hence, we could define (as shown in Figure 10.1) an intermediary zone of management between the end of the useful life (equivalent to the period where maintenance activities cease to occur) and that of the mature life (equivalent to the management of the end-of-life), which could be described as a zone of risks management.

As we can see, this zone of risk management at the tipping point between the useful and mature lives is no different from the broader zone of risk management that any asset endures throughout its useful life. Indeed, the very fact that maintenance activities are usually able to return an asset's reliability to an operationally viable level is enough to prove that risk is being managed on a day-to-day basis. This risk is usually kept under control thanks to the elasticity that characterizes the useful life. As an asset comes closer to the border of its mature life, the maintenance processes that are implemented remain identical but their

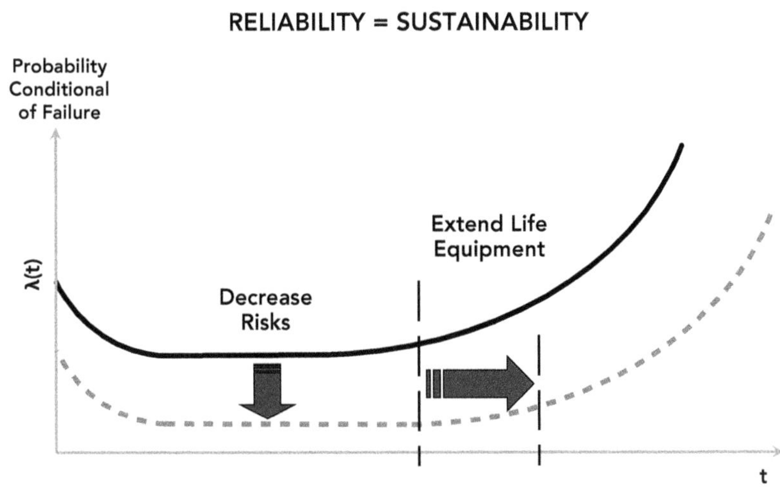

FIGURE 10.1 The levers to manage reliability and operational risks

outcomes are no longer satisfactory. Hence our insistence on the degree of discernment we expect from Asset Managers when dealing with assets that are in the tilting phase between usefulness and maturity; this situation is indeed a complex one that occurs on a daily basis in the industrial field. However, by means of encouragement, we should point out that detecting this transition from one phase to the next does not require any capacity that was not previously channeled throughout the useful life. It only demands that managers embrace a lucid perspective regarding the efficiency of the maintenance activity, which will decrease radically as the asset enters its plastic phase.

Thus, the inefficiency of repairs should be interpreted as a triggering signal: as soon as maintenance actions are no longer able to reduce an asset's failure rate, we can assume that this machine has entered its mature life.

10.3 The Stubbornness of Degradation Kinetics

This first element of decision-making support for maintenance managers brings us to discuss a few misguided but historical beliefs in the fields of operational and maintenance activities that seem urgent to debunk.

The first of these illusory but widespread beliefs states that every time a piece of equipment suffers from a drop in performance, the efforts of the maintenance agent to try and bring it back to its intrinsic reliability level will ensure that the new degradation kinetic (a notion that describes the combination of the speed and the manner in which an asset deteriorates) will be identical to the previous one, *ad vitam aeternam*. Things are not so: in practice, degradation kinetics are changing and evolutive, and on the operational field it is quite easy to demonstrate that equipment tends to behave rather like Figure 10.2 B and not like Figure 10.2 A.

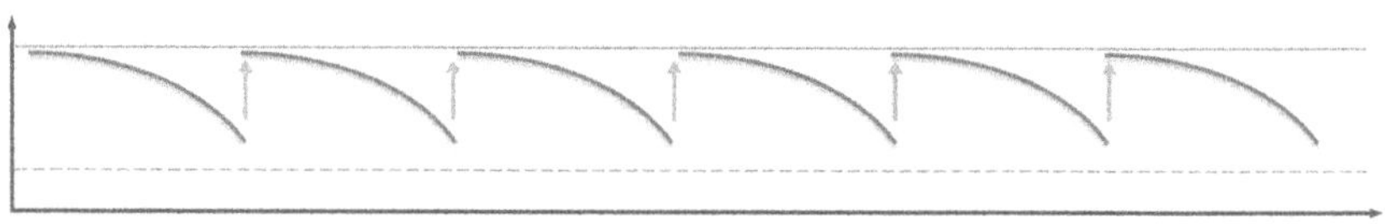

FIGURE A Kinetics of degradation identically reproducible (historical belief)

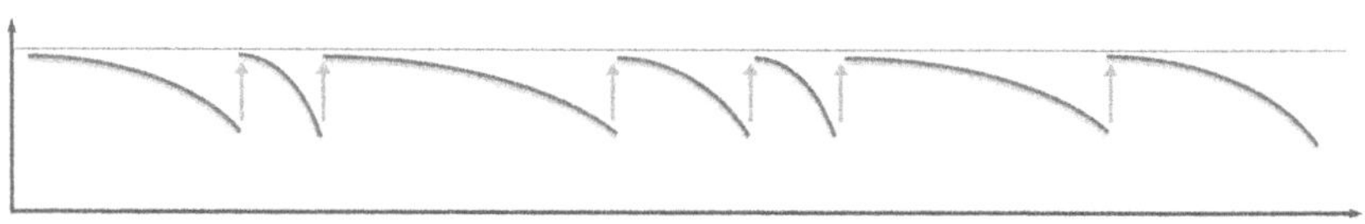

FIGURE B Real kinetics of degradation (plant observations)

FIGURE 10.2 Kinetics of degradation identically reproducible versus real kinetics of degradation

Indeed, the action of repair tends to alter the original integrity of the equipment, which implies that it is unlikely that it should deteriorate in the same fashion with or without human intervention. A ready-made analogy exists within the human body: As we can observe, an organ will never behave in a rigorously identical manner after it has been treated with a surgical intervention.

In line with the ideas that we've just presented, it seems important to deconstruct another historical and erroneous belief. This belief relies on the "*as good as new*" hypothesis, which promotes the idea that assets will systematically return to their previous levels of performance—close to its original performance—if they are still in their useful (and elastic) lives and if the maintenance operation that they've applied was of adequate quality. This hypothesis is so comfortable, from the perspective of maintenance managers, that it tends to obscure the fact that in many cases, it will take another form that we could describe as "*as bad as old.*"

Figure 10.3 illustrates the effect of such a phenomenon, by presenting a superposition of the degradation processes, which in time make the machines less performant "*for good,*" therefore placing it in its mature life.

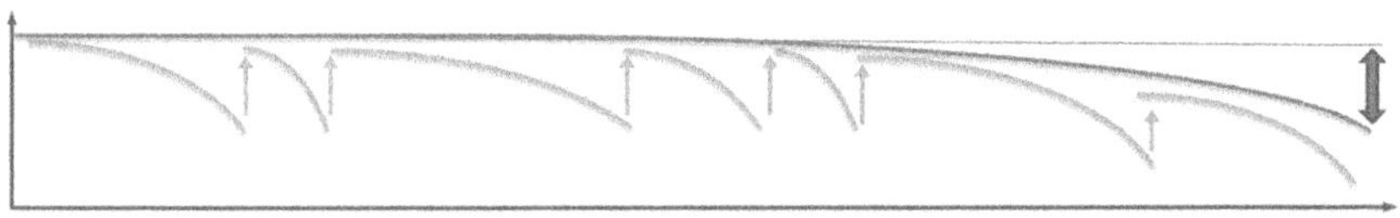

FIGURE 10.3 The "as good as new" hypothesis inevitably slips towards the "as bad as old" hypothesis

If we focus on the matter of costs, we can also gain valuable insights from the historical myths and trends operating in the field of maintenance. This will allow us access to very useful elements of reflection as well as to an improved discernment over the necessary and natural effort aimed at mitigating or even mastering operational costs. One should therefore always ensure that in the context of their legitimate attempts to reduce costs, maintenance managers and their staff be careful not to go beyond the minimal supportable cost (corresponding to the asset's intrinsic reliability level). If cost reductions were to exceed this minimal level, maintenance managers would take the risk of harming the asset's normal performance and function to the extent where the asset would require an additional reliability that would yet be hazardous in the context of an intensive reduction of costs. The asset would therefore be deprived of its functional sustainability. This phenomenon of counter-productive management of one's asset fleet, unfortunately common in the corporate reality, has come to be known among experts as the "decapitalization of assets."

Hence, the key to improving one's discernment over risk reduction operations would be to pinpoint, prior to any cost reduction, a strict limit in order to preserve the asset's elasticity. Otherwise, one will always be exposed to the risk of consuming the assets' potential too brutally, before they even endure natural degradation (Figure 10.4).

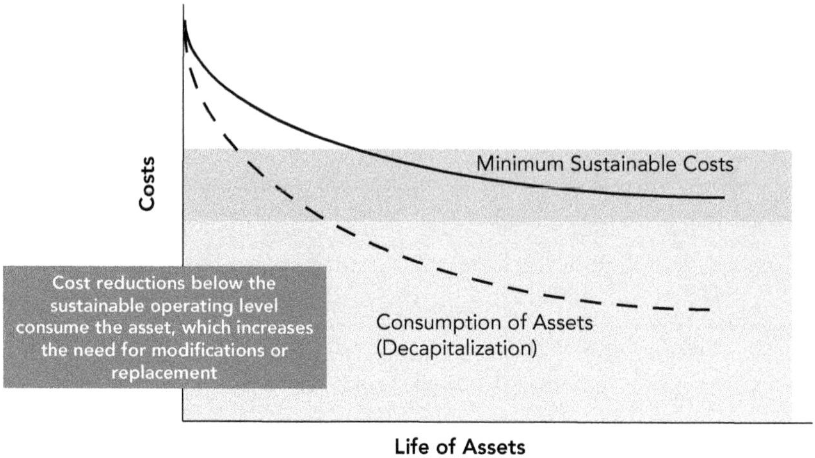

FIGURE 10.4 Decapitalizing assets through uncertain cuts in costs (John S. Mitchell, *Physical Asset Management Handbook*, Ed. 2002)

We will now attempt to show how Asset Management can facilitate the task of the operational and maintenance staffs in the context of the implementation of the decisions made on the basis of the risk/cost trade-offs, particularly in the cases of maintenance or inspection-related interventions.

Figure 10.5 shows a typical modelling of the definition of an optimum, based on the simultaneous exposition of decreasing costs and increasing risks, which allows one to reach an informed decision in a quasi-intuitive manner.

Let us therefore focus on the curves presenting the decreasing costs and the monetized risks and imagine that in the context of a decision-making process, we were to envision various performance scenarios. We would then simply consider base case, worst case, and best case scenarios. Of course, an infinite number of specific cases could arise from these three situations: Figure 10.5 only presents a fragmentary sample of the diversity of the cases that we've already been faced with in our professional experience. After establishing these curves, one should in principle project on the economic ordinate the minimal values of the *best* and *worst* cases. From these two points, one should then trace lines in order to reach the abscissas, which are the graphic representation of the potential decision points. One can thus observe that at least three geometrical figures may be formed within the curves, based on these points and lines:

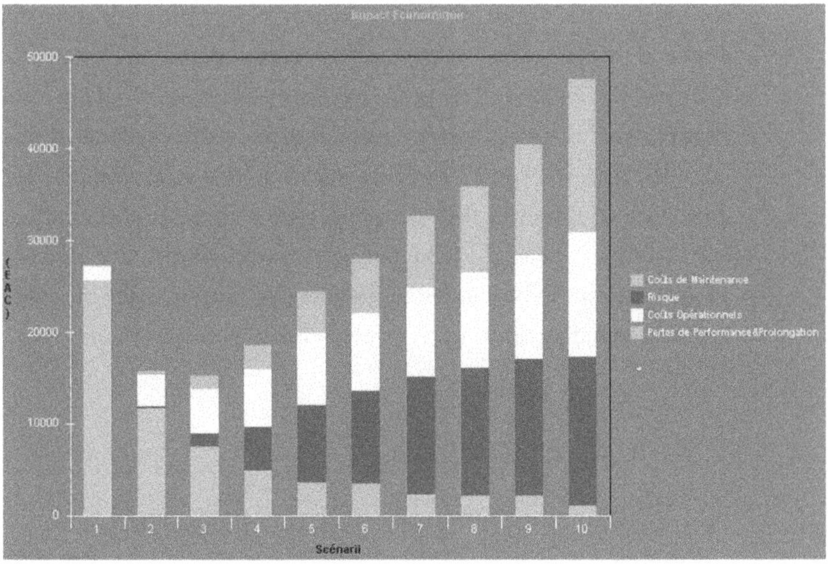

FIGURE 10.5 Representation of monetized costs and risks (Simulator ASSETSIM screenshot)

- **First specific case.** We are faced with a rectangle that has a very wide basis and a very small height. This is the ideal scenario, since any decision made within the frame defined by the rectangle's base, however imperfect, may only have a negligible economic impact. This is known as a *low risk/cost sensibility* decision-making scenario, or as a "recumbent brick."
- **Second specific case.** We are faced with a rectangle that has a very small basis and a tall height. This is undoubtedly the most constraining scenario, since any error in assessment will generate considerable economic consequences. This is known as a *high risk/cost sensibility* decision-making scenario, or as a "standing brick."
- **Third specific case.** We are faced with a figure that is close to a square, and whose basis and height are therefore quasi-equivalent. In this context, it is obviously difficult to elect the optimal decision since we are dealing with economic elasticities analogous to that which we already knew prior to the modelling of this figure. In this situation, the analyst could feel that he has undertaken an unnecessary effort, but we should

observe that there is an efficient stratagem to overcome this apparent dead end. Indeed, when analyses of risk/cost trade-offs do not suffice to provide discriminative decision-making solutions, we should remind ourselves that these analyses were established on the sole basis of two variables: risks and costs. When dealing with quasi-square figures that do not suffice to provide the operational team with an optimal solution, one should encourage the field manager to seek a third and sometimes a fourth variable to be injected in the curve, in order to make the curve assume a "standing brick" or "recumbent brick" position (Figure 10.6).

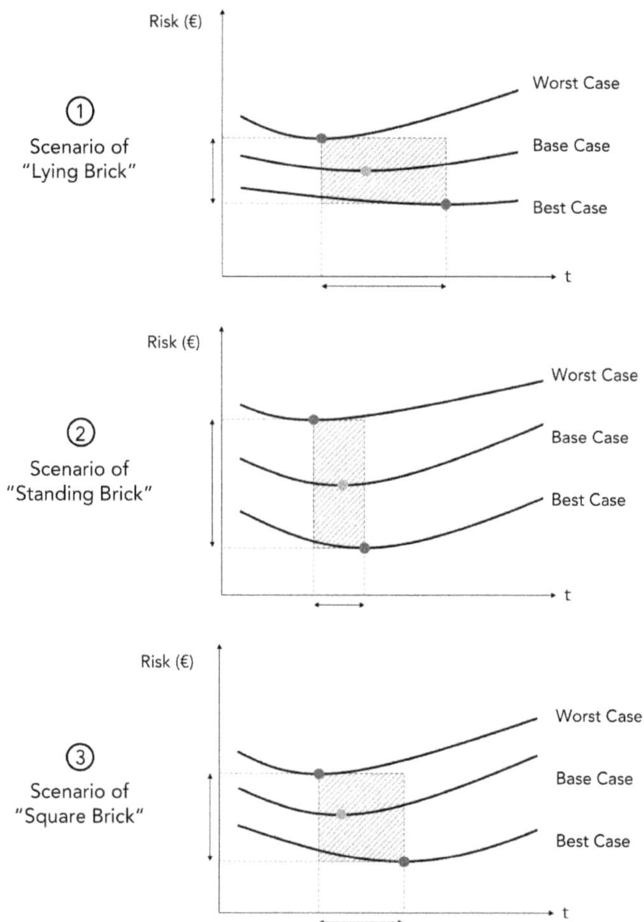

FIGURE 10.6 How "bricks" assist decision-making in Asset Management

Based on hundreds of experimentations, we can confidently assert that the vast majority of decisions made using this method will take the form of "recumbent bricks" and therefore induce a relatively low sensibility in terms of economic consequences. Of course, if this were not the case, it would have been clear even before the emergence of Asset Management, which only brings an objective demonstration to something that the nature of decisions has always provided to the industrial world. Nevertheless, approximately one out of three decisions elaborated in this fashion results in cases of "standing bricks" or of quasi-square figures. We must also keep in mind that in some cases, injecting new variables in order to result in "recumbent brick" scenarios will fail to produce a satisfactory outcome, depending on the robustness of the model relative to the entry variables.

By means of a conclusion, we should wish to insist on the fact that the important phenomenon here is the generalization of the inclusion of a true notion of "measured risk," both in operational terms and in the day-to-day practices of managers. Indeed, too much time has been wasted throughout the course of industrial history relying solely on the two extreme scenarios: either dealing with inconsiderate risks that had not been assessed in the decision-making process or adopting attitudes characterized by a radical aversion to risk that brought managers to assume overly conservative positions. In this regard, the rigorous and practical perspective that we have defined for measured risk may only participate in the development of Asset Management notions and values in the industrial field.

CHAPTER 11

Asset Replacement and Disposal

This chapter will focus on an oft-neglected segment of the life of industrial assets, namely the management of their *end-of-life*. This topic is undoubtedly a vast one, since it is characterized by a massive lack of solutions rather than by a profusion of precise problems.

11.1 Extracting Value from Assets: Until When?

We've already mentioned earlier in this text the prevalence of typical rake-shaped organizations in the industrial world, as well as the deplorable effects induced by this organizational model on the extraction of value and the realization of the corporate objectives, whatever they may be. We have also brought into light an analogy between each "head" of the rake and the various segments of the assets' life cycles; furthermore, we have been able to observe that these segments usually fell under the accountability of a precise manager (procurement manager, maintenance manager, etc.)

However, we can also observe that in most cases the segment that consists of managing the assets' end-of-life phases and therefore in time, in supervising their disposal, is often left without its own dedicated manager. More disturbingly, it is perceived as a "bastard" field within productive organizations (infrastructures and industries) since cases are constructed and conducted (in a quasi-case-by-case assessment) either by the maintenance department when the asset's potential is viewed as low, or by the engineering department (on the other fringe of the asset's life cycle) if the latter has been tasked with managing this particular investment.

And yet it is quite clear that managing the end-of-life implies taking care of assets that, as we've seen, are characterized by performance signals equivalent to those presented by equipment in its mature life. Thus, it is self-evident that the maintenance function (whose goal is to *maintain* and *preserve*) cannot be the most adequately prepared to handle assets whose failure rates are no longer *maintainable.*

Similarly, it seems ill-advised to treat the management of the end-of-life as a fundamentally "new" investment—which would therefore be managed by the engineering branch—since (a) it is by no means a "new" investment and, (b) organizations are equipped with a Performance History of the asset (whatever the quality of these performances) that should be taken into account rather than behaving as though the organization was dealing with a brand-new investment.[1]

Based on these different perspectives, one is inclined to reflect on the following question: why do organizations behave so inadequately in the face of the management of the end-of-life of their assets? Or, as a corollary, why is this segment deprived of having its own dedicated manager, as all other segments in the life cycle do?

Every time I've been brought to ask these questions to the managers of organizations I was working with, I was met with candid surprise: "*Oh, we've never thought about it this way.*" "*I'm afraid that's just the way we do things.*" "*We've never asked this particular question.*" "*The end-of-life has never presented itself to us as a segment in itself.*"

In other words, this question seemed to perplex a vast majority of industry leaders. It would seem that this issue is relatively easy to resolve, but that the main difficulty lies *in its very acceptation.* One must therefore ask the question: why is it so hard to accept that the asset's end-of-life must be managed differently from its useful life? Why won't organizations acknowledge that the treatment demanded by "mature" assets differs from that demanded by "useful" assets?

Part of the answer to these problematics has transpired in our introduction to this chapter: objectively, most of the industrials I converse with do not regard the issue of end-of-life as a serious problem. Yet, let us elect to dig a little deeper in order to bring into light the subjacent rationales that have led this segment to be so poorly equipped until this day.

From the perspective of the general management, it is clear that this line of questioning may appear unseemly. All things considered and from the point of view of corporate leaders, assets are made to fulfill their function and as long as the organizations pay the price for their operation and maintenance, they are entitled to expect that the operational agents ensure that the asset is kept in function for the longest possible time, without facing unexpected issues. However, this posture turns a blind eye to the fact that it is altogether much more profitable to operate cost-effective assets rather than high-longevity assets.

On the opposite end of the organizational model, one must keep in mind that the efforts of operators and maintenance agents are complete with a real professional pride; hence, admitting that an asset has entered its mature phase may sometimes feel like an admission of failure. This implies that to engage a discussion on the management of the mature life, one must be prepared to bring into question the cultural perceptions that agents may have when dealing with the end-of-life.

If machines could talk, they would request a programmed and assisted death, in accordance with their aging condition and the inevitability of their eventual replacement to sustain the company's profitability. However, industry professionals are not always aware of this. They consider that a machine's lifespan can be indefinitely extended. Replacing a gear here, an engine or an axle there will keep it going. They know that, after a while, breakdowns will become more frequent, and production and maintenance costs will increase, but why not keep the machine running since its purchase price has already been absorbed?

"I can see you patting yourself on the back: you have managed to extract record-breaking production levels out of me, the old machine in the plant. Congratulations! Even your bosses have climbed down from their ivory towers, in three-piece suits, to admire your prowess. This shows the good care you've been taking of me in recent years, replacing damaged parts, completing tricky and expensive modifications in order to keep me as fit as a young man.

"Instead, you should be warning the priest: at this rate, I won't make it through the week. Who do you think I am, Methuselah reborn? I hate to disappoint you, but you should know that I won't be around forever, regardless of how many new parts you strap on me. While the parts might be new, my frame is worn out. My skeleton bends under the weight of the years: if the rate doesn't slow, it will fall to dust. When the time comes, there will be nothing left to do other than to find me a replacement.

"I fancied a lovely death. Someone would press a button and I would fall asleep, smoothly and forever. I was hoping to hear: 'Your mission is complete, old friend, a new machine will soon be replacing you and you can rest in peace.' But that is not what happened! I am sentenced to die of exhaustion. When my lights go out, your fancy bosses will wonder if I was worth the investment. When they realize that I am mortal, they will curse my corpse. What a sad way to go out! I should get used to the idea. One has to die someday, and I can take comfort in the fact that mine is close."

This line of reasoning is fairly common in industry. Ask a maintenance professional, anywhere around the world, about the way they perceive their mission. Their answer will generally be along these lines: "To keep machines operating smoothly for as long as possible." Although this is a noble and legitimate pursuit, I propose to tweak it slightly: "To keep machines operating smoothly for as long as it is economically profitable." The difference is huge! As long as we endeavor to keep the machine operating for as long as possible, we will pass up excellent opportunities to increase efficiency and decrease production costs by replacing it. This will have a significant impact on the production and maintenance budgets.

What do the concepts, tools, practices, and methods derived from the Asset Management culture have to offer in terms of novel and truly pertinent insights in the treatment of this issue?

If we were to synthetize the major conceptual inputs of Asset Management, it would be relatively easy to identify a number of strong arguments and valid reasons promoting the modification of our relation to the management of the end-of-life of assets. Let us therefore consider the argument of the "systemic gaze," or in other words, that of the holistic perspective that characterizes the vision of Asset Management. In this context, it is transparent that one cannot continue to promote the notion that assets in their mature lives, which present very different performance signals from those presented by assets in every other segment of their life cycles, do not require a specific form of management. Indeed, these negative signals are impossible to neglect if one considers the totality of the available data.

Let us now focus on another primordial notion: that of the distinction between the useful and mature lives that we've already discussed. It is this differentiation that provides us with the demonstration that the signals emitted by assets in their end-of-life are indeed dissonant.

Finally, let us reflect on the notion of "extraction of value on the totality of the asset's life cycle." Even if we depended on this restrictive definition of Asset Management, it would transpire that one cannot overlook the management of assets in their end-of-life, since it would be dishonest to exclude the mature life from the life cycle that Asset Management has elected to manage.

These are but a few of a vast series of arguments underlining the importance of implementing a cleaner management of this segment.

However, we think it is more effective to demonstrate the robustness of the Asset Management rationale in favor of an improved management of the end-of-life of assets through a recourse to an analogy that we believe is particularly founded. Indeed, this debate is extraordinarily similar to the very contemporary question of the best ways in which to handle climatic change and the environmental transition. Deep down, these problematics depend on a necessity for behavioral transformation (a complex reaction to induce from agents) and on a long-term focus (which is definitely unusual in the industrial world, even more so when a "short-term" behavior can generate harm on the longer course and for which the accountable agent will not necessarily suffer consequences).

The list of these parallelisms is undoubtedly a long one, and we could easily enumerate plenty more arguments that would be completely analogical to the environmental issue. Thus, we can only invite you to consider, whenever you are faced with arguments promoting the protection of the environment and the maintaining of the value that we all extract from the planet, how similar this problematic is to the question of the management of the end-of-life of industrial assets.

In order to address this topic in its most crucial aspect, it is necessary that we consider the creation of value by the organization. Without coming back to the matter of the optimal practices to be developed by management, it seems fitting to compare the noble ideals put forward by the organizations' statements of intent when describing their "vision," their "commitment," and their "mission" on the one hand, and the manner in which these very organizations choose to handle the question of the mature life on the other. In this regard, Asset Management represents a considerable tool for positive transformation since it embodies a manner for organizations to finally walk the talk. This potential for change does not depend on a theoretical discourse, but on the global realization of the veracity of our approach to value: namely, that organizations may only create value on the entirety of their life cycles, whatever else they may pretend in their annual balance sheets. Hence, the real value extracted from the assets can only make sense if it is sustainable and if the management of these assets is inscribed in the long term, including, in this context, in the end-of-life.

In order to offer concrete answers to these questions, we think it would be useful and profitable to introduce at this stage of our outline a number of factors that punctuate the technical and economic lives of industrial assets. Indeed, it is only by grasping precisely the weight of the cross-performances and the important decision-making points that erupt at every stage of the assets' life cycles, and which are presented in Figure 11.1, that one will be prepared to shine a light on the potential solutions that we will discuss thereafter.

This graph presents a life cycle costing (LCC) curve to which we've superposed a curve showing the revenues corresponding to the operation of the asset. The hypothesis illustrated by this superposition of curves is very similar to what we might have obtained had we studied any other industrial asset operated today. Hence, we can observe a first dot, P1, which corresponds to the asset's ROI (return on investment) since if the currency has kept a stable value ,then this point marks the equivalence of ROI and cumulated expenses of investment and operation. We can consequentially observe that uphill from this P1 point, the organization enters a beneficiary phase of operation for this asset and that in view of the shape of the LCC curve, we can expect the beneficiary phase to undergo growth, inflection, and decline until the LCC curve intersects with the revenue curve once more. Hence, the P2 point represents a point known as *zero profitability*. In other

THE CURVE OF LIFE CYCLE COSTING COMPARED TO THAT OF REVENUE

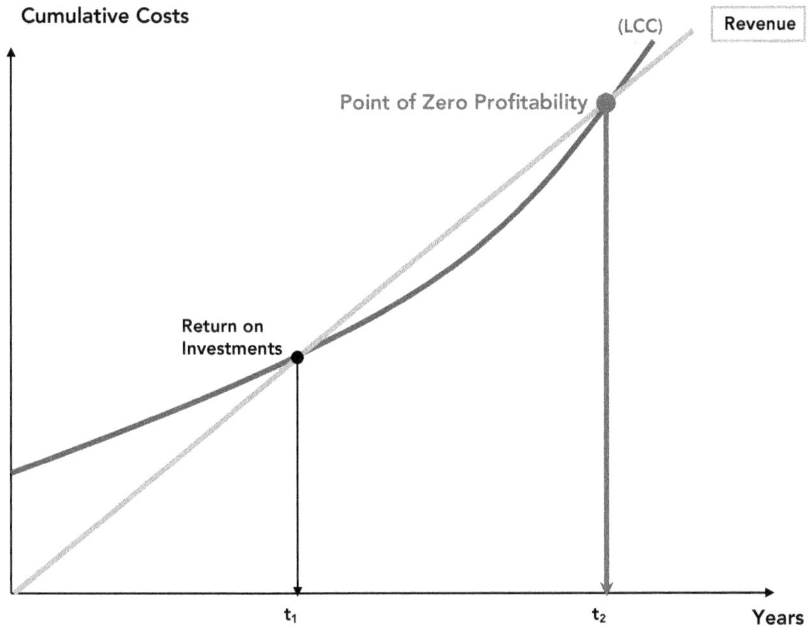

FIGURE 11.1 The life cycle costing curve and its key landmarks

words, it marks the date (which could occur after many years of operation of the asset) at which, despite having fulfilled the respectable mission of generating employment, the cumulative extraction of value from this asset is precisely null. In view of our Asset Management concepts, this curve is therefore not desirable at all for any organization; this point embodies, in fact, the very symbol of an "Anti-Asset Management" model. Worse yet, it is quite clear that if one were to persist in operating an asset beyond this milestone point, one would set foot in a phase of "value destruction" generated by the asset's operation.

In the days when I was still employed as a maintenance engineer, the accountants of the transportation infrastructure organization where I worked spent countless hours teaching me that an industrial asset, as soon as it had paid for itself, became *100% profit.* In later years, and throughout my professional life, I came to hear this assertion repeated time and time again. My role is not to blame the accountants, who had probably never received the engineering instruction that allows one to discern that any capital-intensive industrial equipment is endowed with a technical performance aligned on the notorious "bathtub curve." But as we know, the "bathtub curve" implies the occurrence of the "mature life" and therefore the ineluctable growth of the failure rate towards the end-of-life. One would thus only need to translate this growth into economic terms to result in the demonstration represented by the LCC curve in our figure and to show that analogically, the LCC curve will sooner or later intersect with the revenues curve so that in time, *all* industrial assets will reach a point of zero profitability.

To conclude this chapter in a positive tone, I think we could profit from analyzing what goes on "between" these two pre-evoked milestones (ROI and point of zero profitability) to understand how value extraction is practically wasted throughout an industrial asset's useful life. The fact that the beneficiary phase undergoes a phase of growth, a point of maxima, and a declining phase is easily demonstrable via a recourse to the fundamental principles of marginal economics. These principles teach us how gains or losses are generated by any additional production of goods. In our curve, we can observe that any additional operation of the asset beyond the ROI (return on investment) point generates, at least for some time, an increasing beneficiary margin; we are therefore faced with an operational zone of *positive marginal economy*. As one can derive from this assertion, what will follow is an inflection point, beyond which any additional operation will generate declining margins: We are then faced with an operational zone of *negative marginal economy*.

This approach is a crucial one, since this very observation allowed for one of the major breakthroughs of LCC science in the 1970s: namely, the determination of the economic life of an industrial asset. Thanks to marginal economics, it was then demonstrated that the economic life of an industrial asset corresponds to the operational date where its marginal economy is null. This notion of "economic life" of an asset is in fact rather inaccurate, since it refers to the date at which the profitability of the operation of an industrial asset is at its highest and not to the *period over* which the operation of an asset is most economic.

In fact, what is known as the asset's "economic life" only corresponds to the date during which the asset reached its highest profitability level because at this time the global addition of CAPEX and OPEX, or, in other words, the asset's LCC divided by the number of cumulative operational years is at its lowest point. This is known to accountants and financiers as the Equivalent Annual Cost (EAC). However, at the date of the asset's "economic life," the minimal EAC only embodies an economic advantage for the organization if the latter disposes of the asset at this very date. Should it fail to do so, any additional accumulation of operational time will bring the asset into a zone of negative marginal cost and therefore induce an increase in EAC.

We've deemed it necessary to give a thorough explanation for this notion since all too often, newcomers to our field are uncertain how to address the difference between economic life and opportune time of replacement.

The explanation which we've just operated, relying on LCC concepts, will in time allow us to highlight how thanks to Asset Management practices, we are now able to develop technical and economic decision-support scenarios bearing on the asset's end-of-life. If our demonstration succeeds in facilitating the taking into account of the end-of-life of industrial assets early on in their life cycles and regularly throughout the milestone-punctuated journey that we've already described, it will have been a real success.

11.2 All Assets Are Mortal[2]

In order to enumerate the potential ways in which one could improve the manner in which assets in their end-of-life are dealt with, we will first have to reassert a number of conceptual points. The first of these conceptual reminders will be to bring into question what we mean when we discuss "assets in their end-of-life," since every organization is faced with different types of assets and different types of lives.

Figure 11.2 represents individual assets, which are often subject to replacement since they only have a relatively short lifespan. As soon as they are grouped on the basis of their function (since it has been observed that these assets may have identical purposes and therefore operate the same productive function) they are described as "asset systems." These asset systems are generally endowed with longer lifespans than individual assets. The third section presents what is known as an "asset portfolio," or, in other words, a bundling of asset systems which share a common macro-function within the productive system; undoubtedly, our reader will recognize that one or plural asset portfolios exist within

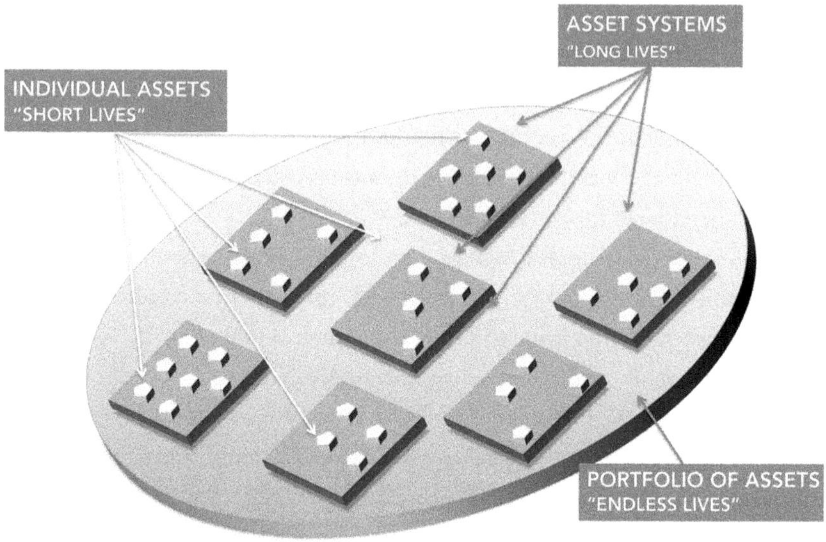

FIGURE 11.2 Various categories of assets according to their various lifespans

his own organization. Asset portfolios boast an *infinite* lifespan; of course, this lifespan is not *intrinsically* infinite, but it can be regarded as such based on the fact that the individual assets and the asset systems whose bundling make up a portfolio are replaced on such a regular basis that it ensures the survival of the portfolio *ad vitae aeternam.*

One should however be wary of confusing this artificial effect of "infinity" with infinity itself: if managers regarded their portfolios and industrial plants as infinite by nature, they could, in a politico-strategic maneuver, cease to treat individual assets and asset systems specifically.

This remark should not be taken lightly, as indeed one frequently hears managers brag, in international conferences and congresses, about the infinite lifespans of their plants or infrastructures (*"Our airport is eight decades old and still works like a charm." "Our refinery was endowed with an infinite lifespan."*) without saying a word about the millions of dollars spent and invested on a daily basis to keep their installations in an acceptable condition of operationality and competitivity.

Now that we've clarified this issue, it seems fitting to evoke the latest methodological trends in the field of end-of-life Asset Management. In this domain as well, a rigorous distinction between the assets' useful and mature lives is essential in order to clearly identify the objects discussed, and more specifically to improve our perception of an organization's asset fleet, which will always be the object of analysis regarding its technical and economic lifespans from a business perspective.

The trade-off techniques that involve risk/cost and CAPEX/OPEX parallels constitute one of the major trends in the analysis methods deployed in the management of the end-of-life. Indeed, the CAPEX/OPEX trade-off allows us to envision scenarios for extending the lifespan of industrial assets.

But the most considerable input in defining the optimal time of replacement of an asset by another asset or by various candidates to its replacement was provided by the emergence of what are now known as "double V" analyses or the "W principle."

The principle of the W-shaped LCC curve presented in Figure 11.3 shows the following phenomenon: The V-shaped curve to the left represents the functioning asset at the time of the analysis. We must keep in mind that all too often within organizations, operational managers and maintenance managers attempt to pinpoint the optimal time of replacement of their end-of-life assets by focusing on the shortcomings inherent to the assets' aging. However, we can demonstrate that the optimal point of replacement of a present asset by a new one is much more strongly determined by the advantages featured on the V-shaped curve to the right of the figure (representing the asset that will replace the current one) rather than by the drawbacks caused by the declining performance and condition of the current asset. Indeed, based on these three variables (the EAC curve of the current asset; that of the future asset; and the time at which the issue of replacing the current asset emerges) we can pinpoint a point, located between the two minimal inflections of the EAC curves, which represents the most opportune moment for abandoning the current asset in favor of the new one.

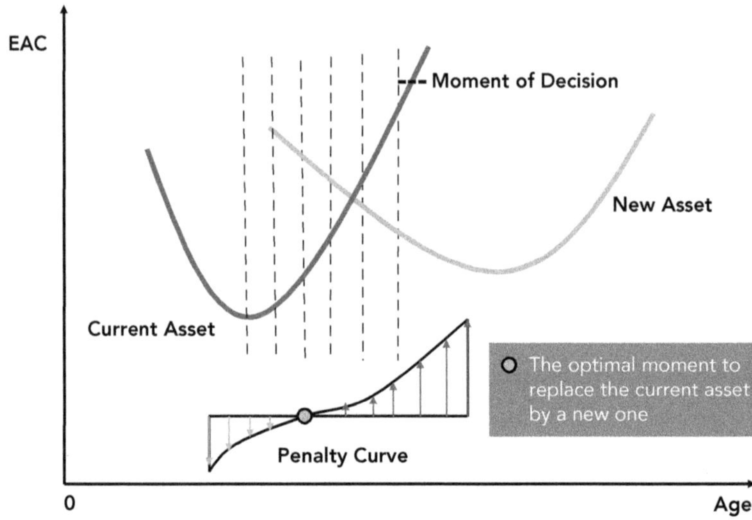

FIGURE 11.3 Principle of the algorithm "W" of Assets Value

This observation emerges from a contradiction between industrial and daily lives, of which I first become aware one morning when a research colleague came into the lab—just as we were pondering the question of "how to modelize this optimal point?"—and shared the followed anecdote with us:

He'd come home tired the previous night and, during dinner with his wife and his two daughters, the main topic of conversation had been the replacement of the family car. He'd therefore had an opportunity to observe that throughout this discussion and debate, his wife and daughters spoke more of the glamorous or technological assets of the prospective car than of the failures and issues of the aging car that they'd been using until then.

This anecdote allowed us to become aware of the fact that focusing solely on the current asset to determine the optimal time of replacement, we were remaining blind to any strategic vision over the technical and economical assets that a machine produced years or decades after the current machine can bring as "added value" for the organization.

Thus, we finally came to the realization that one cannot find every necessary objective and relevant element for the construction of a model of decision-support aiming at value extraction on the sole premise of the studying of operational and end-of-life equipments at a given time.

Later on, we were able to associate to the determination of the optimal point of replacement a penalty curve highlighting the expenses occurred when decisions are made before and after reaching this point. The concept of penalty incrementation allowed us to build an innovative model of investment project ranking for end-of-life assets, as we will describe later on.

The importance of this ranking is explained by the fact that in industrial organizations nowadays, the average volume of investment requests for replacement is equivalent to 3.5 or 4 times the available CAPEX Sustaining budgets.

It is therefore clear that between June and September, when the letters setting the framework for the next year's budget are distributed and collected, every actor involved is able to realize that he cannot be the prioritized recipient, despite his desires. This leads to a strong imbalance between the wishes of Asset Managers and the organizational capacity for CAPEX Sustaining.

Therein lies another major problem, which is often understated by the management leadership. Since the host of project applicants is very large, what ensues is a series of conducts that introduce important biases in the objectivity of project rankings. We will proceed to state the most visible of these biases, while steering clear of any caricature:

- The principle of "decibel management," which consists of granting investment credits to those who "scream the loudest"
- Organized blackmail operations, which usually carry a message such as "if this project is not funded, then I must recuse myself from this business and will not be able to carry out what is expected of me"

We could, of course, state a number of other common practices, which often occur in the organization's working spaces, from favoritism to the expression of more-or-less decent discourses of internal politics.

These practices reveal a true need for objectivity in calculations and consequently a necessity for project ranking in a rationale of Asset Management and coherent leadership. This is the reason why the principle of penalty on both sides of the opportune replacement time for end-of-life assets is so relevant. It allows us today, in a context of plurality of investment request within organizations, to distribute in a multi-year perspective the CAPEX Sustaining credits relative to the penalty incrementation of the candidate projects.

The end-of-life is one of the segments where the largest deficiencies in terms of organizational policies and strategies is observed. We have stated managerial faults, and shortcomings of the implemented prioritization models. Life cycle costing is a boost for the realization of the necessary alignment between the operational field and corporate finance in an Asset Management perspective. Furthermore, it has been demonstrated that every single asset will sooner or later reach its zero profitability point in the long term. Let us also consider that organizations whose cash flow allows them to respond positively to every yearly CAPEX Sustaining project are very rare. Therefore, establishing priorities on objective grounds is a required measure. Multi-year rankings allow for a mitigation of risks and costs relative to the choices in investment projects adapted to the available budget.

Conclusion

As we've previously pointed out, the ultimate aim of Asset Management is to encourage and support, on all levels of industrial organizations, processes of decision-making that seek to extract value from assets on the basis of quantified, thorough and objective information.

Nonetheless, we should observe that companies have not waited on the rise of Asset Management to seek to improve their decision-making: as we know, this desire has always been prevalent within organizational concerns. But what is of a direct concern to us are the ways in which Asset Management can take part in this transition towards top-ranking decision-making.

More specifically yet: our focus is set on the input of elements introduced by Asset Management disciplines in the building up of a proper discernment able to facilitate organizational decision-making in the face of high-stakes cases balancing often contradictory parameters.

Indeed, the very structure of our discipline suffices to demonstrate that decision-making may not be reduced to a mere dimension of speech or to the stroke of a pen; on the contrary, any rational decision depends on an intellectual progression and process.

Throughout this publication, we have strived to illustrate this notion, and to reveal the four slopes that organizations must follow to reach the peak of Asset Management practice. These slopes could be defined as follows:

- The integration and appropriation of methods and techniques relative to trade-offs (risk/cost, CAPEX/OPEX, short/long terms). They are also known as capabilities.
- The development of a necessary base of information and data making it possible to underpin any decision-making scenario for an organization. It is the "IT" dimension of Asset Management, which relies on information technologies and tools.
- The embodiment of decision-making within a strong leadership, which would exercise a mobilizing force over the organization and which would ensure that every segment of the organizational chart is "on the same page."
- The implementation of a solid management system, an essential requisite for the inscription of any Asset Management strategy in the long-term range. This notion is the most holistic out of all four we have stated, in functioning and operational terms.

Learning to make optimal decisions regarding the management of assets: therein lies the true purpose of this publication. However, these decisions may only be reached if the organization also engages in processes of decision-making tied with strategies of management, with operational actions, and of course with the relevant professional actors.

It is only by realizing a synthesis of these various axes that one can acknowledge the full measure of the extraordinary assistance provided by Asset Management in developing a strong leadership within industrial organizations. In practice, a decision is only "acceptable" if someone is able to justify and defend it; neither organizations nor individuals are receptive to that which they do not understand. This is truly the guiding principle of our thinking.

Furthermore, Asset Management is designed to help organizations in the face of their daily functioning but also in their phases of crisis. The crises encountered nowadays throughout the economic world call for immediate answers on the part of the leadership, and these responses are more easily conjured as soon as organizations have learned to devise them holistically.

What does this imply? In short, this calls for the alignment of every organizational strategy, the implementation of a holistic perspective bent on long-term horizons, and the will to break up the "silos" that still divide up organizations.

Of course, this also implies a widespread discussion of risks and of the strategies implemented to confront them: namely, to avoid them, to reduce them, to transfer them, or to consent to a quantified level of exposure. The invaluable strength of Asset Management lies in the choice to make these risks monetizable in an objective manner. Thanks to this notion of calculated risk, our discipline can provide organizations with the opportunity to develop clear processes and informed scenario-building rather than waste time and money in hazardous trials and errors.

Furthermore, it is now clear that no discussion of risks should elude the topic of the widely anachronistic and yet rampant habits which are still widespread in the world of management, *on every level* of the organizations and of the assets' life cycles: design, project management, procurement terms, commissioning, "silos" structures, etc.

In a similar fashion, we've also set out to provide a demystification regarding the issues tied with non-profitable availability (largely due to misallocated CAPEX in the early stages of the life cycle, and superfluous OPEX in the end-of-life) and with the fatalism induced by recurrent functioning breakdowns and failures, which professionals persist in regarding as simultaneously "absurd" and "inevitable" since they seemingly occur "by chance." Our position on this topic is most clear: There is no space for "chances" within an organization that has collectively learned to make objective choices taking into account information regarding risks, costs, and the life cycles of assets.

Through the recourse to this key notion of assets' life cycles, we have been able to introduce essential factors such as the border between useful and mature lives of assets, the elastic and plastic characters of the evolution of material functioning, or the stubbornness of operationals and accountants confronted with these unavoidable facts: that all assets are perishable and that the value one can extract from them is never infinite.

It is thanks to the accumulation of these intuitions transformed into knowledge, and of this knowledge transcribed into methods, that our discipline has been able to take shape. Today, more than ever, Asset Management depends on the production of knowledge, particularly when it comes to the grooming of the professionals of tomorrow. To these professionals-to-be, who I salute and encourage, we will need to repeat relentlessly that the acquisition of a proper discernment in the face of decision-making is vital, but also teach that this discernment is helpless if it is not combined with a will for action and a true sense of urgency.

But anything can be taught, and anything can be learned. With this conviction in mind, I've sought to shed some light over complex recent evolutions, which are still difficult to assess thoroughly. Among these perspectives, one could state the need to recuse the present state of rivalry in order to implement a true complementarity between the roles of maintenance and Asset Management; the necessity to adopt an attitude in favor of proactive information (cf. the "windshield" analogy) rather than to base our analyses exclusively on failure histories and databases (which make up so many "rear mirrors"); or the urgency to distinguish state-of-the-art benchmarking practices from their mere caricatures.

It goes without saying that this publication did not strive to offer an *exhaustive* presentation of the notions and issues that Asset Management responds to. Such a task would indeed be unthinkable, since our discipline inherently eludes any attempt at dogmatism and fixation. On the contrary, we hope and believe that this work will succeed in opening infinite perspectives, and in inviting the reader to engage in a questioning and an enhancement of the reflections touched upon throughout these insights.

Hence, and without any pretense at misplaced humility, we could state that our greatest hope for this essay is that it may soon be outdated.

Notes

Chapter 1

1. Indeed, the assets are scrapped or destroyed on the spot.
2. The object of ISO 55010: 2018.

Chapter 11

1. It may be a deliberate strategic choice to consider a "zero base" replacement, but it is neither a necessity nor a widespread practice.
2. A wink to the spirit of Simone de Beauvoir's reference book, *All Men Are Deadly* (Gallimard, 1947).

Bibliography

Bazerman (Max H), Moore (Don A.), *Judgment in Managerial Decision Making*, 8th edition, ed. Wiley, 2012.

Bazovsky (Igor), *Reliability Theory and Practice*, Dover Publications Inc., 2004.

Burns (Penny), *Strategic Asset Management*, editor of AMQ International, 2012.

Commission de normalisation de gestion d'actifs d'AFNOR, Guide thématique de la gestion d'actifs, IFRAMI, 2017.

Deadman (Clive), *Strategic Asset Management: The Quest for Utility Excellence*, Troubador Publishing, 2010.

De Azevedo (Celso), *Guide de la gestion d'actifs industriels—Si les machines parlaient*, Dunod, 2009.

De Azevedo (Celso), "Optimizing the timely moment of technical-economic replacement of end-of-life assets," *Assetsman*, 2001.

De Azevedo (Celso), "FEEDMAN: l'application des méthodes d'Asset Management en phase de conception d'installations industrielles," *Assetsman*, 2008.

De Azevedo (Celso), Riddell (Harry), *Distinction Between Useful and Mature Life of Assets: Applying Whole Life Cycle Costing to Surface Installations*, Spe applied technology workshop—Field Management of Mature Reservoirs, Scheveningen, The Netherlands 4–7 June , 2002.

ISO 55000:2014 Asset management—Overview, principles and terminology, ISO Ed., Geneva, 2014.

Gestion d'actifs—Aperçu général, principes et terminologie, AFNOR, 2014.

Kervern (G.-Y), Boulenger (P.), *Cindyniques, concepts et mode d'emploi*, Édition Economica, 2007.

Labadessa (Gino), *Front-End Engineering Design : FEED e os Elementos Criticos de Sucesso em Projetos de Capital*, 2014.

Miles (Lawrence D.), *Techniques of Value Analysis and Engineering*, Kindle Edition, 3rd. edition, 1989.

PAS 1192-2:2013—Specification for information management for the capital/delivery phase of construction projects using building information modelling, British Standard Institute, 2013.

Pfeffer (J.), Sutton (R.), "Hard Facts, Dangerous Half-Truths, and Total Nonsense: Profiting from Evidence-Based Management," *Harvard Business Review Press*, 2006.

Scheldon (M. R.), *Life Cycle Costing: A Better Method of Government Procurement*, Westview Press, 1979.

Sunstein (Cass R.), *Choosing Not to Choose: Understanding the Value of Choice*, Oxford University Press, 2015.

Tichy (Noel M.), Bennis (Warren G.), *Judgment: How Winning Leaders Make Great Calls*, 2007.

Walker (Stephen), "Asset Management and Long-Term Planning: Learning from Audit Plannings 2015 to 2017," *Audit New Zealand*, 2017.

Figure Index

Figure Index

Index

R

www.ingramcontent.com/pod-product-compliance
Ingram Content Group UK Ltd.
Pitfield, Milton Keynes, MK11 3LW, UK
UKHW021012120926
8556IPUK00039B/1041

9 780831 136444